Cascade Use in Technologies 2018

EBOOK INSIDE

Die Zugangsinformationen zum eBook Inside finden Sie am Ende des Buchs.

Alexandra Pehlken · Matthias Kalverkamp
Rikka Wittstock
Editors

Cascade Use in Technologies 2018

Internationale Konferenz zur Kaskadennutzung und Kreislaufwirtschaft – Oldenburg 2018

Springer Vieweg

Editors
Alexandra Pehlken
Cascade Use Research Group
Carl von Ossietzky University of Oldenburg
Oldenburg, Germany

Matthias Kalverkamp
Cascade Use Research Group
Carl von Ossietzky University of Oldenburg
Oldenburg, Germany

Rikka Wittstock
Deparment of Accounting and Information Systems
University of Osnabrück
Osnabrück, Germany

ISBN 978-3-662-58608-2 ISBN 978-3-662-57886-5 (eBook)
https://doi.org/10.1007/978-3-662-57886-5

Springer Vieweg

Softcover re-print of the Hardcover 1st edition 2019

This Springer Vieweg imprint is published by the registered company Springer-Verlag GmbH, DE part of Springer Nature
The registered company address is: Heidelberger Platz 3, 14197 Berlin, Germany

Preface

Cascade Use in Technologies is dedicated to cascading approaches of materials and products and fits perfectly to the Circular Economy discussion within Europe and the rest of the world. Most methods used in this context can be more or less allocated to life cycle management approaches with a rather technical perspective, as most of the articles presented here in this book are engineering driven. Therefore, it is a novel forum for reporting technological breakthroughs in the area of cascading materials and products in the domains of Automotive, Electronics, Computing, Renewable Energy Technology and Resource Criticality in general.

Cascading generally addresses the full utilization of components through cascades of reuse, recycling, and recovery (final sink). Process paths are not fixed, as inputs can go directly to recycling or recovery. However, in many cases it is desirable from an environmental perspective to maintain products at the reuse or recycling levels of the cascade. Cascade utilization is a widely applied concept within the biomass domain and the Cascade Use research group fosters the transfer of this concept to other products and materials (e.g. critical and strategic metals). In addition, the generation of ideas on how existing management approaches (e.g. waste management, production planning and control), the given landscape of end-of-life management software and databases on end-of-life products (e.g. company or network specific databases) can effectively support the implementation of the cascade use concept has been identified as an essential towards a circular economy. Waste flows could be minimised and nearly avoided if more knowledge was available on qualitative and quantitative data of products and material flows. All involved stakeholders could manage their supply chains according to the available resources if they knew the time and composition of materials in products.

The cascading approach clearly addresses the Circular Economy goals. By reusing and recycling materials and products, we keep secondary resources in the loop and protect primary reserves in both open and closed loop life cycles. Ideally, the circular economy achieves closed-loop recycling to avoid waste flows directly during manufacturing. In today's reality, however, this is clearly not happening. This fact is a main driver for the creation of this book, the purpose of which is to address challenges and opportunities for implementing the cascade use concept. Why are loops often hindered and what are the barriers and possible options for closing the cycle? The different chapters of this book discuss major challenges and opportunities of recycling/reuse loops within product life cycles, not only focussing on technical approaches but also on economic and logistical aspects. The "International Conference on Cascade Use and Circular Economy - ICCCE 2018" held in Oldenburg i.O., Germany in 2018 brought together experts from around the world and was organized by the Cascade Use research group, funded by the German Federal Ministry of Education and Research (BMBF) under the research grant FONA – Globaler Wandel. After five years of sustainability research in sustainable resource management at Oldenburg University, the Cascade Use research group under the lead of Alexandra Pehlken combined their excellent network to gather most of their partners in the one spot, where the group started: in Oldenburg in the North-western region of Lower Saxony, Germany. Since Cascade Use already had a strong relation to the Shanghai Jiao Tong University, the conference was scientifically supported by Prof. Chen Ming and the AARTI – Alliance of Automotive Recovery Technology Innovation from China.

Contributions from China, Northern America and Europe with case studies focussing on various products and technologies complete this book on "Cascade Use in Technologies". We acknowledge all authors, without whom no book would have been possible in the first place. In addition, we thank our ICCCE 2018 team members for their efforts in organizing and supporting the conference and this book, including the professionals from Springer.

A special thank goes to the German Ministry BMBF, the Federal Ministry of Education and Research for providing the funding for the Cascade Use research group, funded from 2014 to 2019 at the Carl von Ossietzky University Oldenburg. Cascade Use was promoted by the FONA program and guided by the DLR during the past five years and we are grateful for the financial support making the ICCCE 2018 and this book possible.

Enjoy this book as we have enjoyed selecting the papers for this book. Hopefully, Cascade Use in Technologies will help foster new and innovative research worldwide.

The Editors,
Alexandra Pehlken
Matthias Kalverkamp
Rikka Wittstock

Contents

Contributors

Martin Adrion School of Engineering, Pforzheim University, Pforzheim, Germany

Henning Albers Department of Civil and Environmental Engineering, University of Applied Sciences Bremen, Bremen, Germany

Manuel Baumann Institute for Technology Assessment and Systems Analysis (ITAS), Karlsruhe Institute of Technology (KIT), Karlsruhe, Germany

Nicolas Bognar Chair of Sustainable Manufacturing & Life Cycle Engineering, Institute of Machine Tools and Production Technology (IWF), Battery LabFactory Braunschweig (BLB), Technische Universität Braunschweig, Braunschweig, Germany

Urte Brand Institute of Networked Energy Systems, DLR, Oldenburg, Germany

Felipe Cerdas Chair of Sustainable Manufacturing & Life Cycle Engineering, Institute of Machine Tools and Production Technology (IWF), Battery LabFactory Braunschweig (BLB), Technische Universität Braunschweig, Braunschweig, Germany

Sascha Diedler Institute of Environmental Technology and Energy Economics, Hamburg University of Technology, Hamburg, Germany

Frauke Germer Department of Civil and Environmental Engineering, University of Applied Sciences Bremen, Bremen, Germany

Juan Camillo Gomez Institute of Networked Energy Systems, DLR, Oldenburg, Germany

Thomas Guidat Department of Machine Tools and Factory Management, Technical University Berlin, Berlin, Germany

Weiqun Han College of Science & Arts, Jianghan University, Hubei, China; Hubei Key Laboratory of Advanced Technology for Automotive Components and Hubei Collaborative Innovation Center for Automotive Components Technology, Wuhan University of Technology, Hubei, China

Christoph Herrmann Chair of Sustainable Manufacturing & Life Cycle Engineering, Institute of Machine Tools and Production Technology (IWF), Battery LabFactory Braunschweig (BLB), Technische Universität Braunschweig, Braunschweig, Germany

Merle Heyken Department of Ecological Economics, Carl von Ossietzky University of Oldenburg, Oldenburg, Germany

Matthias Kalverkamp Cascade Use Research Group, Carl von Ossietzky University of Oldenburg, Oldenburg, Germany

Ersin Karadeniz School of Engineering, Pforzheim University, Pforzheim, Germany

Neele Karbe Cascade Use Research Group, Carl von Ossietzky University of Oldenburg, Oldenburg, Germany

Muztoba Ahmad Khan Industrial and Management Systems Engineering, West Virginia University, Morgantown, WV, USA

Christian Klinke School of Engineering, Pforzheim University, Pforzheim, Germany

Björn Koch Cascade Use Research Group, Carl von Ossietzky University of Oldenburg, Oldenburg, Germany

Kerstin Kuchta Institute of Environmental Technology and Energy Economics, Hamburg University of Technology, Hamburg, Germany

Mehmet Ali Kucuker Institute of Environmental Technology and Energy Economics, Hamburg University of Technology, Hamburg, Germany

Zhu Lingyun School of Mechanical Engineering, Shanghai Jiao Tong University, Shanghai, P.R. China

Mark Mennenga Chair of Sustainable Manufacturing & Life Cycle Engineering, Institute of Machine Tools and Production Technology (IWF), Battery LabFactory Braunschweig (BLB), Technische Universität Braunschweig, Braunschweig, Germany

Chen Ming School of Mechanical Engineering, Shanghai Jiao Tong University, Shanghai, P.R. China

Alexandra Pehlken Cascade Use Research Group, Carl von Ossietzky University, Oldenburg, Germany

Fernando Peñaherrera Cascade Use, Carl von Ossietzky University of Oldenburg, Oldenburg, Germany

Jens F. Peters Helmholtz Institute Ulm (HIU), Karlsruhe Institute of Technology (KIT), Karlsruhe, Germany

Julian Rickert Chair of Sustainable Manufacturing & Life Cycle Engineering, Institute of Machine Tools and Production Technology (IWF), Battery LabFactory Braunschweig (BLB), Technische Universität Braunschweig, Braunschweig, Germany

Yuan Shi Hubei Key Laboratory of Advanced Technology for Automotive Components and Hubei Collaborative Innovation Center for Automotive Components Technology, Wuhan University of Technology, Hubei, China

Pang-Chieh Sui Hubei Key Laboratory of Advanced Technology for Automotive Components and Hubei Collaborative Innovation Center for Automotive Components Technology, Wuhan University of Technology, Hubei, China

Katharina Szczepaniak Technische Universität Hamburg, Hamburg, Germany

Nina Tsydenova Department of Ecological Economics, Carl von Ossietzky University of Oldenburg, Oldenburg, Germany

Thomas Vogt Institute of Networked Energy Systems, DLR, Oldenburg, Germany

Michael Wark Department of Chemistry, Carl von Ossietzky University of Oldenburg, Oldenburg, Germany

Marcel Weil Helmholtz Institute Ulm (HIU), Institute for Technology Assessment and Systems Analysis (ITAS), Karlsruhe Institute of Technology (KIT), Karlsruhe, Germany

Aleksandra Patrycja Wewer Department of Machine Tools and Factory Management, Technical University Berlin, Berlin, Germany

Rikka Wittstock Department of Accounting and Information Systems, University of Osnabrück, Osnabrück, Germany

Jörg Woidasky School of Engineering, Pforzheim University, Pforzheim, Germany

Thorsten Wuest Industrial and Management Systems Engineering, West Virginia University, Morgantown, WV, USA

Kalle Wulf Department of Civil and Environmental Engineering, University of Applied Sciences Bremen, Bremen, Germany

Jinsheng Xiao Hubei Key Laboratory of Advanced Technology for Automotive Components and Hubei Collaborative Innovation Center for Automotive Components Technology, Wuhan University of Technology, Hubei, China

Xiaochen Xu Institute of Environmental Technology and Energy Economics, Hamburg University of Technology, Hamburg, Germany

Goufang Zhang Hubei Key Laboratory of Advanced Technology for Automotive Components and Hubei Collaborative Innovation Center for Automotive Components Technology, Wuhan University of Technology, Hubei, China

Cascade Utilization During the End-of-Life of Product Service Systems: Synergies and Challenges

Muztoba Ahmad Khan[1], Matthias Kalverkamp[2] and Thorsten Wuest[1]
[1]Industrial and Management Systems Engineering, West Virginia University, Morgantown, WV 4001, USA
mdkhan@mix.wvu.edu, thwuest@mail.wvu.edu
[2]Cascade Use Research Group, Carl von Ossietzky University of Oldenburg, Oldenburg, 26111, Germany
matthias.kalverkamp@uni-oldenburg.de

Abstract

The circular economy concept is receiving increasing attention from academia and businesses as a conceivable means to decouple economic growth from material consumption. Product Service Systems (PSS), primarily due to their sustainability potential, have been identified as a promising lever that can facilitate the transition towards a circular economy. However, a product may not be more resource efficient or have reduced environmental impacts just because it is marketed through one of the various PSS business models. In this regard, the comprehensive End-of-Life (EOL) management of PSS can play a crucial role by maximizing the utilization of a product's remaining value. In this paper, we consider the applicability of the cascade use methodology proposed by Kalverkamp et al. [15] in the context of PSS. Additionally, we explore the possible synergies and associated challenges between PSS business models and cascade utilization.

1 Introduction

Rapid technological advancement, changing consumer preferences and increasing market competition prompt manufacturing companies to introduce a variety of often reduced lifespan products at a cheaper price. Consequently, a majority of manufacturers is trying to increase their profits based on sales volume and overall cost reduction. However, because of innovation through new knowledge, changes in consumers' perceived needs and excessive costs of repair or maintenance, these products quickly become obsolete [1] even before the end of their actual physical lifetime or economic value [2]. The current socioeconomic system is primarily based on a linear production and consumption model following a 'take-make-use-dispose' philosophy. This only elevates the environmental and economic challenges caused by inefficient use of scarce resources.

The concern about this growing environmental load related to economic growth has prompted increased interest in alternative ways of achieving more sustainable economic models through enhanced resource efficiency. In this context, the circular economy concept is receiving increasing attention from academia and businesses as a conceivable means to decouple economic growth from material consumption [3]. Unlike linear production and consumption models, the circular economy promotes greater resource productivity by reducing waste and use of virgin materials through reuse, remanufacturing, and recycling of End-of-Life (EOL) products [4-6]. However, from an economic prospective, circular business models need to assume more business risks compared to linear business models. This is primarily due to the complexity associated with reverse supply chains and the uncertain economic value of EOL products [7].

In this regard, Product-Service Systems (PSS) have been widely recognized as a promising lever that can support and facilitate the transition towards a circular economy [4, 6, 8-9]. The PSS business model, in which the use or the function of a product is sold instead of the product's ownership, can mitigate the provider's risk to maintain circularity of EOL products, components, or materials. One of the main reasons is that in case of such PSS, providers have access and better control of the products as well as information regarding the condition, quantity and timing of the returns [10]. Additionally, companies that retain ownership of a product are responsible for the whole lifecycle of their product [11] and thus will have an intrinsic motivation to (i) prolong the useful life span of their products, (ii) maximize utilization, (iii) ensure energy and material efficiency, as well as (iv) reuse products, components, and materials as much as possible after the end of the product's life [9, 12-14].

However, a product may not be more resource efficient [9] or have reduced environmental impacts [14] just because it is marketed through one of the different PSS business models. In this respect, the comprehensive EOL management of PSS that will help to maximize the utilization of product's remaining value can play a crucial role. The cascade use methodology, which has been widely utilized in the biomass domain, offers a broader perspective on the EOL [15]. Recently, Kalverkamp et al. applied this methodology in the context of products that are not marketed through renting or sharing (i.e., PSS) business models [15]. In this paper, we consider the applicability of the cascade use methodology in the context of PSS, while exploring possible synergies and associated challenges between PSS business models and cascade utilization.

The remainder of the paper is structured as follows. In the next section, key concepts for the later discussion are presented in the form of an overview of circular economy, PSS and cascade use. The third section summarizes the cascade use methodology proposed by Kalverkamp et al. [15] and considers its applicability in the context of PSS. The

A. Pehlken et al. (Eds.), *Cascade Use in Technologies 2018*,
https://doi.org/10.1007/978-3-662-57886-5_1

fourth section sheds light on the potential synergies and associated challenges between PSS business models and cascade utilization. The final section summarizes and concludes the paper and provides future research directions.

2 Key Concepts

In this section, key concepts for the later discussion are presented in the form of an overview of circular economy, PSS and cascade use.

2.1 Circular Economy

The circular economy concept is attracting significant attention from researchers, industry and policy makers. In the literature, the most common and widely accepted definition of circular economy is given by the Ellen MacArthur Foundation [6], which defines circular economy as a "system restorative and regenerative by design, which aims to maintain products, components and materials at their highest utility and value". The Ellen MacArthur Foundation has also identified the following three circular economy value drivers: (i) pursuing resource efficiency, (ii) extending the lifespan of products and (iii) closing the material loop. The circular economy puts forward the idea of restoration and circularity in order to decouple the environmental burden from economic growth by enabling multiple closed-loop cycles of reuse, remanufacturing, and recycling [23-24]. The circular economy business model has a strong connection with PSS due to the fact that the responsibility related to the management of EOL product's lifecycles shifts to the PSS provider. This shift of responsibility from a private person (user) to a professional entity (PSS provider) supports the sustainable management of closed-loop industrial systems where materials are recollected, reused, remanufactured, and recycled [25].

2.2 Product Service Systems (PSS)

PSS can be understood as a special case of servitization - a concept introduced by Vandermerwe and Rada in 1988 [16], long before PSS were introduced. Servitization describes the phenomenon of manufacturing firms developing value propositions by incorporating additional services in order to attain a competitive edge in the market [17]. Servitization and PSS, both describe the same concept (i.e., "a marketable set of products and services" [1]) but PSS usually involves the sustainability context in addition to the somewhat 'economic only' context of servitization. In 1999, Goedkoop et al. [18] introduced the term PSS and defined it as "system of products, services, networks of players and supporting infrastructure that continuously strives to be competitive, satisfy customer needs and have a lower environmental impact than traditional business models". A solution composed of physical products and related services may be harder to replicate for a competitor, compared to solely product and process-based manufacturing [19]. Additionally, integrated services mean more satisfactory experience for customers and generally increased revenue for the manufacturers [20].
PSS business models can be categorized into the following three distinct types [17, 21]:

- *Product oriented business models:* The primary purpose of this business model is to provide tangible value to the customer. The ownership of the product is transferred to the customer, while the PSS provider sells additional services in the form of maintenance, upgrades, or EOL take-back.
- *Use oriented business models:* In this business model, the PSS provider sells the use or availability of a product that is not owned by the customer. Examples of this type are product leasing or sharing.
- *Result oriented business models:* In this business model, the PSS provider sells a result or capability of a product and not ownership. For example, instead of selling a printer to a customer, the company can sell the result, such as document management capability.

This paper mainly focuses on the use and result oriented PSS business models. A business strategy built around these business models establishes a value proposition in which manufacturers retain the product ownership and are responsible for its functionality, maintenance, upgrade, and EOL management. This transfer of the responsibility to manufacturer creates an incentive for them to design best possible products in terms of superior functionality, reduced operational (i.e., less consumables inputs) and maintenance cost, and better reusability, re-manufacturability, and recyclability [22]. As a result, PSS may prove to be a more resource efficient and effective solution with less environmental impact compared to conventional product-oriented solutions.

2.3 Cascade Use

The cascade use methodology originates from the forestry sector and has been widely utilized in the biomass domain. Cascade use can be defined as the efficient utilization of resources by using a certain resource sequentially for different purposes [26-27]. The objective is to first exploit the products, components, and materials on higher cascade levels for a longer period of time, before using them as an energy source. In the case of products, adopting a cascade use means pre-planning and designing the route of the EOL products from one product or component to another [28], for example, using EOL electric vehicle batteries in stationary applications before recycling the materials. Another common example from the biomass domain is the use of solid timber in higher value products with large dimensions, instead of chipping it or using it as fuel for energy [29].

Circular economy and cascading utilization have many parallels and similarities [30] in that both strategies promise to increase resource efficiency by a circular management and multiple uses of resources [31]. However, cascade use primarily focuses on the utilization possibilities (e.g., reuse, remanufacture and recycling) of a particular resource, while circular economy provides a more holistic approach [30].

3 Cascade Methodology and its Applicability in the Context of PSS

In the next section, we first summarize the cascade use methodology proposed by Kalverkamp et al. [15], and then consider the applicability of this methodology in the context of PSS.

3.1 Cascade Use Methodology

The cascade use methodology aims to highlight and integrate the complexity of end-of-life options into the management of product lifecycles. The cascade use methodology shows how products and eventually materials cascade through reuse and recycling to recovery (ideally avoiding landfill). Originally, the concept of cascade use stems from the biomass domain where it represents how renewable resources such as wood, i.e. wood fiber, cascade through consecutive processes of use, reuse, and recycling before being treated as an energy source [32]. The term "cascade" or "cascade use" (and similar versions) is also used outside the biomass domain, for example in the context of lifecycle management, reuse, and product returns [33].
For the cascade use methodology, the biomass-domain "cascade" serves as a blueprint in combination with the steps 'reuse', 'recycle' and 'recovery'. These steps are derived from the waste management hierarchy [34]. Fig. 1 shows the cascade use methodology and depicts clearly the increasing complexity and variety of end-of-life options at the levels reuse, recycle, and recovery. This cascade perspective neglects the possibility that products and materials can as well move from a lower to a higher cascade level. An example how this might be facilitated is through an upcycling or reuse process, transforming a discarded product into another (new) product. In such case, the product/material flow exits the cascade and enters another cascade, representing the end-of-life of this new product. Products and materials can also remain at the same cascade level through iterations of reuse (e.g., remanufacturing) or recycling (e.g., up-/down-cycling). This cascade does not consider landfill as landfill does not contribute to circularity [34].

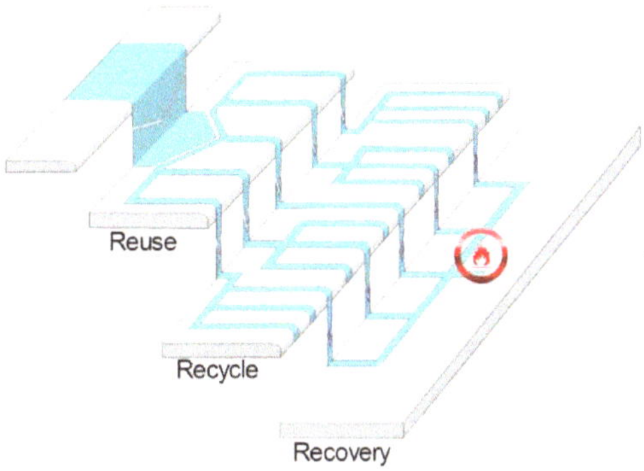

Figure 1. Cascade Use Methodology [15]

Although the cascade perspective influences the product lifecycle from its beginning, the reality at the product's end-of-life may not fit the planned lifecycle even if circular business models such as PSS or closed-loop supply chains were used to manage the product lifecycle. At some point, products may leave the system "unscheduled". For example, when third-parties take advantage of products offered to the "outside system". The latter case is most likely to occur in systems associated with a transfer of ownership (rather than PSS). However, even products used in a PSS setup may eventually end up with another owner. At this point, the end-of-life-options increase and so does the complexity of managing the EOL. The visualization of different cascades fosters alternative end-of-life solutions, in supporting decision makers to identify economic and environmental potential in the different 'streams' of the cascade by integrating market realities (e.g. trade with used products) of changing end-of-life options. The cascade use methodology acknowledges this complexity and recognizes that one supply chain owner can hardly manage all potential end-of-life scenarios.
Kalverkamp et al. [15] introduced a case study on the cascade use of tires (Fig. 2). It was found that after their first use, tires are reused without any alteration in another market, which tolerates a lower tread depth of tires. In the next cascade level, tires with too low tread depth are retreaded as a remanufacturing operation. Before using tires in the energy recovery cascade level, the tires are recycled into new products such as shoe soles or artificial turfs.

Figure 2. Cascade Use of Tires [15]

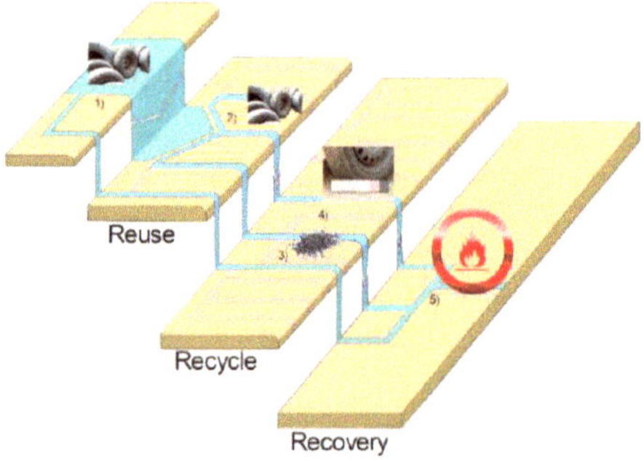

In a second case study, the authors outlined how a product component moves through the different cascade levels while not being managed in a typical closed-loop supply chain scenario. They identify third parties taking advantage of discarded suspension control arms for remanufacturing purposes. Their case shows that the third party closes the component loop and improves the environmental impact of the component through an extended component lifetime. The related process innovation for the remanufacturing of this component highlights that third parties, not being part of a closed-loop supply chain or a PSS, can contribute significantly to the sustainability of products and components. This marks one of the potential challenges of PSS where due to different reasons the product producer may be no longer interested in a product life extension although such could still provide environmental benefits.

3.2 Applicability of the Proposed Methodology in the Context of PSS

Instead of assuming integrated reverse flows from the EOL towards predefined reuse and recycling processes, the proposed methodology takes into consideration that highly integrated, totally market oriented and intermediate organizational types coexist. As a result, it includes both open and closed-loop supply chain perspectives to reflect the broad variety and complexity of different EOL options.

On the other hand, use/result oriented PSS by definition form some kind of closed-loop supply chain as the providers retain ownership of the PSS. The closed-loops are especially prevalent at the reuse and remanufacturing levels. However, at the recycling and recovery levels, a closed-loop setting may not be a viable option for a majority of the PSS providers. For example, the lack of economies of scale may prevent a washing machine provider to recycle all the materials of an obsolete washing machine when building a new one. For this reason, PSS providers may feel more comfortable in an open loop setting at the recycling and recovery levels.

In this regard, the proposed methodology could be a good fit for PSS, given their need for both open and closed-loop supply chain perspectives. Therefore, we propose to consider the cascade use methodology in the comprehensive EOL management of PSS to support the transition towards a circular economy.

4 Synergies and Challenges of PSS Business Models and Cascade Utilization

The authors of the proposed methodology identified policy, new technology (e.g., for remanufacturing or recycling), business models and raw material prices as the possible influencing factors that affect the mass flows within different cascades. However, the authors do not consider the aspect that there could be several factors that may influence the eco-efficiency of a particular cascade level. For example, there are several factors that influence the eco-efficiency [35] of remanufacturing, such as product design, build quality and information regarding the condition, quantity, and timing of the returns. Similarly, different business models with their distinct characteristics can be understood as one of such influencing factors that can play an important role in determining the efficiency within a cascade level. This observation calls for an exploration of the possible synergies and associated challenges between PSS business models and cascade utilization.

4.1 Possible Synergies

Compared to traditional manufacturers who rely on product sales, PSS providers have better capabilities (systematic recovery, product condition and usage data, technical knowledge, investment potential, market for recovered products etc.) and economic incentives to ensure the optimal reuse and remanufacture of EOL products. Consequently, a product marketed through a PSS business model may have a better chance to be utilized at the higher cascade levels such as reuse and remanufacture due to readily available information throughout the supply chain. Retained ownership of the products, closer customer relations and the information regarding product location help PSS providers with systematic recovery through a better organization of the collection (for remanufacture/recycle) and reallocation (for reuse) activities during EOL or when the subscription ends [7, 23]. As the EOL products originate from various sources and different working conditions, they do not show uniform quality conditions, which results in unique remanufacturing needs [36]. However, due to advances in information technology, PSS providers can gather large amounts of usage data throughout the lifecycle and track the condition of the products and components in real-time, allowing for less uncertainty and better performance of remanufacturing activities [23]. Furthermore, in order to restore a product to original or better condition, remanufacturing requires considerable expertise and knowledge of the product that is generally readily available to PSS providers [7, 37]. Another challenge for remanufactured or second-hand products is that their price needs to be adjusted in order to attract customers. However, in a PSS setting, the acceptance and demand for those products is significantly improved as the customers only use the product without having its ownership transferred (i.e., paying for it) [38].

Product design can also play a crucial role to ensure higher levels of cascading use such as reuse, remanufacturing and recycling while delaying final sinks such as energy recovery or landfill. Thinking in the perspective of cascade use will help to appropriately consider and reflect the EOL requirements during the design phase. Certain design decisions taken at the Begin-Of-Life (BOL) stage can have major implications in the management of EOL products. For example, it can be very inefficient to manage an EOL product if its design makes it difficult to disassemble in order to remove hazardous substances [39]. On the other side, the design can reflect the future need for easy disassembly, e.g., by using threaded fasteners instead of epoxies, however, this is often associated with an increase in cost. Since the PSS providers retain the ownership of the product, the entire lifecycle management of the product becomes their responsibility and they tend to focus more on the Total Cost of Ownership (TCO) instead of mainly focusing on design and manufacturing associated cost. Therefore, PSS providers have several incentives to design more durable and flexible products while keeping different eco-design criteria, such as design for disassembly, remanufacturing, and recycling, in mind [40]. Moreover, designing for technological cycle [41] enables alternative revenue models that create value from waste [42], for example, 'cradle-to-cradle' [43] – where raw materials are chosen based on their recyclability nature; and 'industrial symbiosis' [44], where by-products from one process become feedstock for another process.

4.2 Associated Challenges

One of the challenges of maintaining a closed-loop supply chain within PSS business models is that it may limit the innovation potential of third party remanufacturers, who can often lead the way towards the remanufacturing of new parts that are not being remanufactured by the OEMs [15]. In some cases, third party remanufacturers can even create a remanufactured component that performs better than the equivalent new part manufactured by the OEM. For example, a third-party remanufacturer claims to have overcome the common failure of a throttle body by improving its design [45]. These kind of solutions bring economic gains for both the remanufacturer (profit) and customer (lower cost) while contributing towards the environment by delaying material recycling [15]. However, in a closed-loop setting the flow of EOL products and components may not reach such third-party remanufacturers, resulting in a loss of innovation opportunity.

Due to the complexity at the EOL, it is very difficult for PSS providers to manage all available EOL options on their own. Consequently, several stakeholders govern the decision on product design and EOL recovery option. There are several factors that affect these decisions; these factors are related to engineering, business, environmental, and societal aspects [46]. However, there can be conflict between these factors, which may result in varying prospective between stakeholders. For example, material recycler's interest in pure and easy to recycle material may not coincide with PSS provider's interest in composite light weight material [46].

Another drawback specific to PSS business models is that the users may not use the PSS in the recommended way since they do not actually own the PSS. This may adversely affect the useful lifespan and thus the underlying objective of cascade use. The cost of maintaining reverse logistics can be another important concern for the PSS providers. Even when a particular remanufacturing or recycling operation is technically feasible, the costs of recovery operations must be less than the recovered value in order to make remanufacturing or recycling economically attractive for PSS providers [47].

5 Conclusions and Future Research

This paper investigated the applicability of the cascade use methodology in the context of PSS. It further explored possible synergies and associated challenges related to PSS business models and cascade utilization. Cascade use methodology includes both open and closed-loop supply chain perspectives to reflect the broad variety and complexity of different EOL options. Thinking in the perspective of cascades will enable PSS providers to consider a broad variety of EOL alternatives in addition to the originally planned options in the closed-loop setting. Consequently, the environmental and economic benefits of the PSS business models may go beyond the initially designed lifecycles. PSS have a better chance to be utilized at the higher cascade levels due to readily available lifecycle information. Furthermore, PSS providers have more incentives compared to traditional manufacturers to design a product with the objective of retaining them at higher cascade levels for a longer time.

Future research should concentrate on how PSS providers can go beyond their closed-loop supply chain setting and incorporate solutions offered by third party remanufacturers and third markets. Furthermore, future research should develop a method that will consider the collective interests of all stakeholders when designing a PSS and deciding on EOL recovery options in order to transcend the boundaries of individual stakeholders. A methodology to assess the cascading degree within a PSS supply chain can play an important role in proper implementation of the circular economy. Lastly, a wide implementation of PSS business models and therefore a transition towards the circular economy will require simultaneous support from manufacturers, customers, policy makers, lawyers, and regulatory institutions.

6 Zusammenfassung

Das Konzept der Kreislaufwirtschaft steht als mögliche Maßnahme zur Entkopplung des wirtschaftlichen Wachstums von einem steigenden Materialverbrauch zunehmend im Fokus von Wissenschaft und Wirtschaft. Aufgrund ihrer Nachhaltigkeitspotenziale wird Produkt-Service-Systemen (PSS) eine vielversprechende Rolle bei der Umstellung zu einer Kreislaufwirtschaft zugesprochen. Allein die Tatsache, dass ein Produkt als Teil eines der vielfältigen PSS Geschäftsmodelle vermarktet wird, bedeutet jedoch nicht dass es als ressourceneffizient angesehen werden kann oder geringere Umweltwirkungen verursacht. Das umfassende End-of-Life (EOL) Management von PSS kann in diesem Sinne eine wesentliche Rolle spielen, indem es die Ausnutzung des verbleibenden Produktwerts maximiert. In diesem Beitrag wird untersucht, inwiefern sich das von Kalverkamp et al. [15] vorgeschlagene Konzept der Kaskadennutzung auf PSS übertragen lässt. Zudem werden mögliche Synergien und Herausforderungen zwischen PSS Geschäftsmodellen und Kaskadennutzung betrachtet.

Acknowledgments. The authors thank West Virginia University and the J. Wayne and Kathy Faculty Fellowship in Engineering for the support. Matthias Kalverkamp was financially supported by the German Federal Ministry of Education and Research (BMBF) in the Globaler Wandel Research Scheme (Grant No. 01LN1310A).

7 References

[1] Cooper, T. (2004). Inadequate life? Evidence of consumer attitudes to product obsolescence. Journal of Consumer Policy, 27(4), 421-449.

[2] Umeda, Y., Daimon, T., & Kondoh, S. (2005). Proposal of Decision Support Method for Life Cycle Strategy by Estimating Value and Physical Lifetimes—Case Study—. In Environmentally Conscious Design and Inverse Manufacturing, 2005. Eco Design 2005. Fourth International Symposium on (pp. 606-613). IEEE.

[3] Sauvé, S., Bernard, S., & Sloan, P. (2016). Environmental sciences, sustainable development and circular economy: Alternative concepts for trans-disciplinary research. Environmental Development, 17, 48-56.

[4] Elia V, Gnoni MG, Tornese F. Measuring circular economy strategies through index methods: A critical analysis. J Clean Prod 2017;142(4):2741- 2751.

[5] Elia, V., Gnoni, M. G., & Tornese, F. (2017). Measuring circular economy strategies through index methods: A critical analysis. Journal of Cleaner Production, 142, 2741-2751.

[6] MacArthur, E. (2013). Towards the Circular Economy, Economic and Business Rationale for an Accelerated Transition. Ellen MacArthur Foundation: Cowes, UK.

[7] Linder, M., & Williander, M. (2017). Circular business model innovation: inherent uncertainties. Business Strategy and the Environment, 26(2), 182-196.

[8] Lewandowski, M. (2016). Designing the business models for circular economy—Towards the conceptual framework. Sustainability, 8(1), 43.

[9] Tukker, A. (2015). Product services for a resource-efficient and circular economy–a review. Journal of cleaner production, 97, 76-91.

[10] Chierici, E., & Copani, G. (2016). Remanufacturing with Upgrade PSS for New Sustainable Business Models. Procedia CIRP, 47, 531-536.

[11] Shehab, E. M., & Roy, R. (2006). Product service-systems: issues and challenges. In the Fourth International Conference on Manufacturing Research (ICMR 2006). John Moores University, Liverpool, 5th–7th September.

[12] Michelini, G., Moraes, R. N., Cunha, R. N., Costa, J. M., & Ometto, A. R. (2017). From linear to circular economy: PSS conducting the transition. Procedia CIRP, 64, 2-6.

[13] Guidat, T., Barquet, A. P., Widera, H., Rozenfeld, H., & Seliger, G. (2014). Guidelines for the definition of innovative industrial product-service systems (PSS) business models for remanufacturing. Procedia CIRP, 16, 193-198.

[14] Kjaer, L. L., Pagoropoulos, A., Schmidt, J. H., & McAloone, T. C. (2016). Challenges when evaluating product/service-systems through life cycle assessment. Journal of Cleaner Production, 120, 95-104.

[15] Kalverkamp, M., Pehlken, A., & Wuest, T. (2017). Cascade Use and the Management of Product Lifecycles. Sustainability, 9(9), 1540.

[16] Vandermerwe, S., & Rada, J. (1988). Servitization of Business: Adding Value by Adding Services. European Management Journal, 6(4), 314-324, DOI 10.1016/0263-2373(88)90033-3

[17] Baines, T. S., Lightfoot, H. W., Evans, S., Neely, A., Greenough, R., Peppard, J., ... & Alcock, J. R. (2007). State-of-the-art in product-service systems. Proceedings of the Institution of Mechanical Engineers, Part B: journal of engineering manufacture, 221(10), 1543-1552.

[18] Goedkoop, Mark. (1999). Product Service systems, Ecological and Economic Basics.
[19] Martinez, V., Bastl, M., Kingston, J., & Evans, S. (2010). Challenges in transforming manufacturing organisations into product-service providers. Journal of manufacturing technology management, 21(4), 449-469.
[20] Annarelli, A., Battistella, C., & Nonino, F. (2016). Product service system: A conceptual framework from a systematic review. Journal of Cleaner Production, 139, 1011–1032. https://doi.org/10.1016/j.jclepro.2016.08.061
[21] Tukker, A. (2004). Eight Types of Product-Service System: Eight Ways to Sustainability? Experiences from SusProNet, 260, 246-260, DOI: 10.1002/bse.414
[22] Vaittinen, E. (2014). Business model innovation in product-service systems. Tampere University of Technology.
[23] Bressanelli, G., Adrodegari, F., Perona, M., & Saccani, N. (2018). Exploring How Usage-Focused Business Models Enable Circular Economy through Digital Technologies. Sustainability, 10(3), 639.
[24] Braungart, M., McDonough, W., & Bollinger, A. (2007). Cradle-to-cradle design: creating healthy emissions–a strategy for eco-effective product and system design. Journal of cleaner production, 15(13-14), 1337-1348.
[25] Catulli, M., & Dodourova, M. (2013, January). Innovation for a Circular Economy: Exploring the Product Service Systems Concept. In ISPIM Conference Proceedings (p. 1). The International Society for Professional Innovation Management (ISPIM).
[26] Vis, M., Mantau, U., & Allen, B. (2016). Study on the optimised cascading use of wood. No 394/PP/ENT/RCH/14/7689, Final report, vol. 2016. European Commission, Brussels, p. 337.
[27] Höglmeier, K., Weber-Blaschke, G., & Richter, K. (2013). Potentials for cascading of recovered wood from building deconstruction—A case study for south-east Germany. Resources, Conservation and Recycling, 78, 81-91.
[28] Vezzoli, C., & Manzini, E. (2008). Design for environmental sustainability (p. 4). London: Springer.
[29] Husgafvel, R., Linkosalmi, L., Hughes, M., Kanerva, J., & Dahl, O. (2017). Forest sector circular economy development in Finland: A regional study on sustainability driven competitive advantage and an assessment of the potential for cascading recovered solid wood. Journal of Cleaner Production.
[30] Mair, C., & Stern, T. (2017). Cascading Utilization of Wood: a Matter of Circular Economy?. Current Forestry Reports, 3(4), 281-295.
[31] Bezama, A. (2016). Let us discuss how cascading can help implement the circular economy and the bio-economy strategies. Waste Manag. Res. 34, 593e594. http://dx.doi.org/10.1177/0734242X16657973.
[32 Haberl, H., & Geissler, S. (2000). Cascade utilization of biomass: strategies for a more efficient use of a scarce resource. Ecological Engineering, 16, 111-121. DOI: 10.1016/S0925-8574(00)00059-8.
[33] Guide, V. D. R., & Wassenhove, L. N. (2006). Closed□loop supply chains: an introduction to the feature issue (part 1). Production and Operations Management, 15(3), 345-350. DOI: 10.1111/j.1937-5956.2006.tb00249.x
[34] European Commission, Closing the loop - An EU action plan for the Circular Economy. COM (2015) 614 final. [Online] Available: http://eur-lex.europa.eu/legal-content/EN/TXT/?uri=CELEX:52015DC0614. Accessed on: Jul. 05, 2016.
[35] Deng, Q., Liu, X., & Liao, H. (2015). Identifying critical factors in the eco-efficiency of remanufacturing based on the fuzzy DEMATEL method. Sustainability, 7(11), 15527-15547.
[36] Joshi, A. D., Gupta, S. M., & Ishigaki, A. (2017). Evaluation of design alternatives of sensor embedded end-of-life products in multiple periods. Procedia CIRP, 61, 98-103.
[37] Pearce, J. A. (2009). The profit-making allure of product reconstruction. MIT Sloan management review, 50(3), 59.
[38] Guidat, T., Barquet, A. P., Widera, H., Rozenfeld, H., & Seliger, G. (2014). Guidelines for the definition of innovative industrial product-service systems (PSS) business models for remanufacturing. Procedia CIRP, 16, 193-198.
[39] Lee, H. M., Sundin, E., & Nasr, N. (2012). Review of end-of-life management issues in sustainable electronic products. In Sustainable Manufacturing (pp. 119-124). Springer, Berlin, Heidelberg.
[40] Pourabdollahian, G., & Copani, G. (2016). Toward development of PSS-oriented business models for micro-manufacturing. Procedia CIRP, 47, 507-512.
[41] Romero, D., & Rossi, M. (2017). Towards circular lean product-service systems. Procedia CIRP, 64, 13-18.
[42] Bocken, N. M., Short, S. W., Rana, P., & Evans, S. (2014). A literature and practice review to develop sustainable business model archetypes. Journal of cleaner production, 65, 42-56.
[43] McDonough, W., & Braungart, M. (2010). Cradle to cradle: Remaking the way we make things. North point press.
[44] Lombardi, D. R., & Laybourn, P. (2012). Redefining industrial symbiosis. Journal of Industrial Ecology, 16(1), 28-37.
[45] ACtronics. Volvo—Magneti Marelli Gasklephuis. 2016. Available online: https://www.actronics.eu/volvo-magneti-marelli/ (accessed on 06 June 2018).
[46] Ziout, A., Azab, A., & Atwan, M. (2014). A holistic approach for decision on selection of end-of-life products recovery options. Journal of cleaner production, 65, 497-516.
[47] Guide Jr, V. D. R., & Van Wassenhove, L. N. (2009). The evolution of closed-loop supply chain research. Operations research, 57(1), 10-18.

Shifting Remanufactured Products from Used to New

Aleksandra Patrycja Wewer[1] and Thomas Guidat[1]
[1]Department of Machine Tools and Factory Management, Technical University Berlin, Berlin, 10587, Germany,
wewer@mf.tu-berlin.de

Abstract

Humans use more resources than the earth can regenerate and the demand on materials is expected to rise further. The worldwide economy follows mainly the linear rules, in which only one life cycle of a product is considered. Remanufacturing, the reuse of products after their treatment, even if achieving a turnover of round 30 billion Euros in Europe accounts only for a small proportion of the manufacturing. In this paper the integration of remanufacturing in the production of new units is introduced in order to enable a shift of remanufactured products from used to new ones. The opportunity for this approach is explained based on the example of household appliances.

1 Introduction

Humans strive to lead a better life. The definition of a better life depends on the individual itself, their living situation and the recognized possibilities. Globalization supports not only the exchange of goods and information but also desires. The inner need to own luxury products as for example a car, a television or a smartphone can be observed all over the globe [1, 2]. A higher consumption of goods results in a higher resource consumption. In a linear economy raw materials are extracted, processed, transformed to products, used and discarded. In recent years, it has become common to collect and to return some materials to the production site [3]. Nevertheless, the majority of the economy is still linear. Circle Economy calculated the global circularity metric, which describes the proportion between the cycled materials and the material input, to be 9.1 percent in 2015 [4]. Additionally, humans worldwide use more resources than the earth can regenerate as the earth overshoot day was already on the 2nd of August [5]. Furthermore, the world population is steadily growing as well as the challenge to ensure an acceptable living standard [6]. To return the material to the production seems not to be sufficient to bridge the gap. The reuse of products is discussed and implemented in several industries as a further part of the solution. The products are treated and offered as a further purchase option often for a lower price. This additionally addresses another customer group with a lower willingness to pay and thus can lead to a further consumption and resource use [7]. Returning complete components or assemblies to a production of new units to save both material and work needed is not addressed yet.

2 Influence of the Ownership on Remanufacturing

To manufacture a product, materials, energy and work have to be added. Even if the product itself loses value over time, it still retains the material value as well as the value of production that can be used in reprocessing and thus lead to energy and material savings. The possible energy savings amount to up to 85 percent compared to the production of a new product [8], but this has to be checked for each case individually [9]. There are three main approaches to treat a product after the previous use phase: no treatment, repair and remanufacturing. The repair consists only of actions to restore the functionality of the product. In contrast to this, remanufacturing aims to achieve a product condition that is at least as good as new. According to BS8887: Part 2, remanufacturing is an industrial process of "returning a used product to at least its original performance with a warranty that is equivalent to or better than that of the newly manufactured product". During this process the products go through the steps of inspection, disassembly, repair and replacement, cleaning and reassembly [10-12]. Remanufacturing is the highest level of treatment, since it guarantees at least the same quality, functionality, safety and warranty as a new product, even though it does not address explicitly the appearance and the perception of the product.

Several researchers have tried to explore the extent of remanufacturing in the economy. In the USA, in 2012 the United States Trade Commission introduced the report Remanufactured Goods: An Overview of the U.S. and Global industries, Markets, and Trade [13]. In 2015, the European Remanufacturing Network released a Remanufacturing Market Study [14]. In 2017, a report of VDI Centre for Resource Efficiency was published [15]. According to these data, the remanufacturing industry is worth round 30 billion Euros and employs around 190,000 people across Europe [14]. Nevertheless, it reaches only two percent of the turnover from production [15]. The share of various industrial sectors in remanufacturing is comparable in both the USA and Europe. The aviation industry with more than 12 billion Euros turnover from remanufacturing in Europe, the automotive industry with more than 7 billion Euros, and the Heavy Duty and Off-Road Equipment industry with more than 4 billion Euros turnover are the three main sectors of remanufacturing in Europe and together achieve more than 80 percent of the whole turnover [15]. In the mentioned

A. Pehlken et al. (Eds.), *Cascade Use in Technologies 2018*,
https://doi.org/10.1007/978-3-662-57886-5_2

industries, the manufactured products can be acquired through high initial investment and therefore a longer use through maintenance is intended. Further, there are legislative regulations, such as e.g. the requirements for maintenance in the aviation industry that enforce a regular treatment and enable a continuous availability of old parts.
The availability of cores as well as the reverse logistic are mentioned as one of the most important challenges of remanufacturing [16].The occurrence of the end of life due to a functionality failure or the general dissatisfaction with the product cannot be predicted precisely and depend amongst others on the usage intensity, the quality of the product and the willingness to purchase a new product by the user. The ownership of a product determines to some extent the unit responsible for the return. If ownership is with the user and no other requirements must be met, the date and the end of life strategy are assigned to them. If ownership remains with a party responsible for the end of life treatment, due to a business model providing a service of usability and not the ownership itself, a so-called Product- Service- System, the return date can be defined in advance. A common example is a leasing contract on a car, where the exact date and the condition of the product at the return are predetermined [17]. The challenge of the reverse logistic to handle the uncertainty of date and quantity of the return can be reduced and the product can be supplied to the process. The Figure 1 illustrates four exemplary ownership models under consideration of remanufacturing.

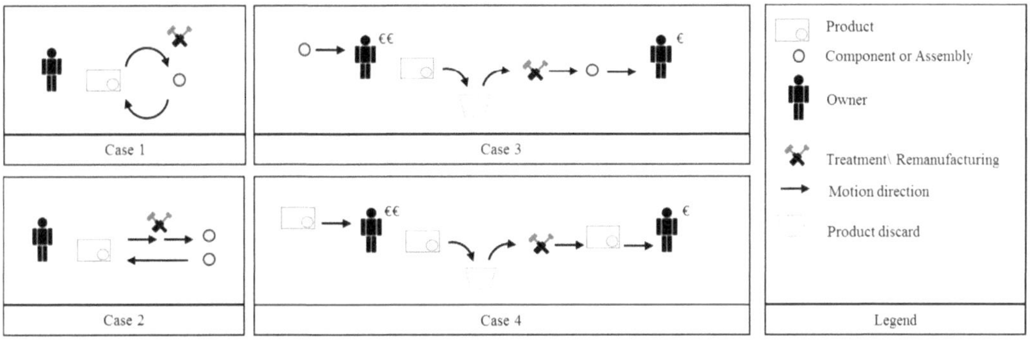

Figure 1. Ownership during remanufacturing

In case 1 the ownership remains with the customer all the time. A customer owns a product as well as all included assemblies and components. If an assembly has to be changed, this assembly can be removed, treated and reinstalled. In practice, automotive gears are remanufactured after a request of the owner and reinstalled in the same product. There is a direct connection to the owner and the observed errors can be queried. The spare part is usually adapted to a specific car model and can be exchanged only by an equivalent part.
In case 2 the ownership of the product remains with the customer, but the ownership of the assembly changes. A customer owns the product as well as all included assemblies and components. If an assembly has to be changed, another, already treated assembly is installed. The removed assembly is treated and stocked by the remanufacturer. Alternators are the most common examples, as they are installed in several car models and are available in large quantities. The exchange can be done as promptly as when using a new spare part.
In case 3 the ownership of the product remains with the customer, but the ownership of the assembly changes. A customer owns a product as well as all included assemblies and components. If an assembly has to be changed, a new assembly is installed. The removed assembly is discarded by the first owner and is bought by a remanufacturing company. It is treated and sold to another customer requiring a more favourable price-performance ratio.
Case 4 is similar to case 3 with the difference that ownership refers to the whole product. An ascending trend of this can be observed for electronic devices, such as tablets or laptops, which no longer meet the requirements of the first user but have a performance sufficient for other customer groups.
The described cases are applied in different industry sectors and business models and address a customer group with a lower willingness to pay [18-21]. Remanufactured products are perceived as used products, the customer therefore attributes a higher failure risk to them [20-22]. In several cases, where the product is used directly by the customer, as for example clothes or smartphones, additional information stating that the product was treated can lead to a worse perception of the product. To analyse the influence of information provided on the perception of used products, Ackermann et al. conducted a study in which the attitudes towards new, briefly worn and worn plus certified as equal to new pairs of pants were elaborated. The results of the study show that the respondents have worse feelings about the worn and certified as equal to new pair of pants than the just worn ones [22].
Using a different method of remanufacturing and simultaneously providing less information to the customer may shift remanufactured products from used to new ones without changing the customer attitude and therefore address the same customer group as for new products and with the same willingness to pay. In this paper a further model for remanufacturing is introduced and illustrated in Figure 2. In the introduced case, the customer acts according to the principles of the linear economy, i.e. buy, use, discard. The discarded product is not landfilled or recycled but is disassembled, and the assemblies and components are remanufactured. The remanufactured parts are not used as spare

parts but as parts for a production of new units. The new product contains new and remanufactured components. The proportion of the remanufactured components differs in each end product. The main difference to the current remanufacturing approach is that additionally to the functionality aspect visual aspects are addressed, as the produced product shows no signs of usage. For this approach the knowledge about the location and availability of cores is crucial. Therefore, ownership of the product and thus the possibility of deciding when a product is to be taken back should remain with the manufacturer. From an ecological point of view, this approach shows the advantage that no new customer group is addressed and therefore no additional resource consumption is expected.

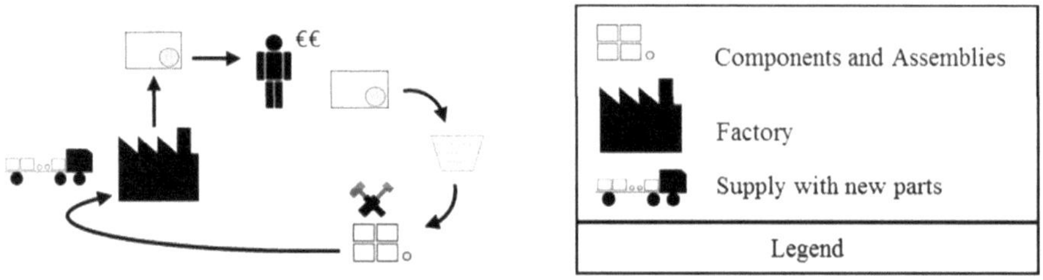

Figure 2. Integration of remanufacturing in a new production

3 The Potential of Household Appliances for Remanufacturing

Domestic appliances have become an integral part of European households. More than 95 percent of all households in Germany have e.g. a washing machine. According to the findings of Stamminger and Goerdeler 4.5 washes per week are made on average in a German household [24]. The additional effort of handwashing is no longer considered acceptable.

The average life span of a washing machine is difficult to determine. An online survey shows that the majority of washing machines are used longer than ten years [25]. If the long failure-free life time of washing machines is considered, it seems that there is no potential for remanufacturing in this sector. During such a long life span, several new models with additional features appear on the market. The cost for a single spare part amounts to a significant portion of the cost of a new washing machine. It often seems uneconomical to repair these devices after the warranty has expired, which is why a new product is bought. On the other hand, it is common to repair used washing machines in small repair shops where the spare parts are extracted from other used machines and to sell them for a small portion of the cost of a new one. An industrial process with high quality requirements would cost more money and shrink the price difference.

If a closer look at the construction of a washing machine is taken, two groups of components can be differentiated: Components that can be seen by the customer, such as the drum or the housing, and components within the machine, such as the pump. Prior research conducted by the author at the spare parts dealers has revealed that some components of the second group used by a major German household appliances manufacturer are installed in different models of a washing machine for more than twenty years. If the usage phase of a washing machine of ten years is sought, remanufactured components could be supplied to the production after half of the production timespan. Additionally, components taken from broken machines due to transportation damages or warranty claims could be used earlier.

The German washing machine market is saturated. Therefore, there is no need to attract a new customer group with a lower willingness to pay. Conventionally remanufactured washing machines would compete with low cost washing machines. Integrating remanufacturing in the production of new machines, however, would address the same customer group and reduce resource consumption without awaking the perception of a higher functionality risk.

Even if the market is saturated, there is a high demand for washing machines. It is assumed that an average household remains for 60 years, i.e. the time span between the start of independent living at 20 years and death at the age of 80. These figures represent a rough orientation rather than statistical data. With the average life time of a washing machine estimated at 15 years, four washing machines are needed for one household. With approximately 40 million households, 2.7 million washing machines are needed each year. This assumption corresponds with the market data. In 2011, 2.9 million washing machines were sold in Germany [26]. A further advantage for the market and a disadvantage for the environment is the fact that the relative cost of durable goods in Europe is sinking. For example, the necessary worktime for a washing machine, i.e. the time a worker with an average salary has to work to earn enough money to buy a product at an average price, amounted to 225 hours in 1960. In 2007, an average washing machine was achievable after 36 hours [28]. This can foster the willingness to buy new products instead of using them as long as they work without a defect. Using remanufactured components and assemblies would lessen the environmental impact of this approach without limiting the customer.

In addition, there are still markets with a high demand for low cost washing machines. In Vietnam in 2012, only 22.7 percent of all households owned a washing machine. The demand for them is increasing, as is the number of households [27]. To meet the demand, products based on older technology can be produced locally and therefore extend the time a component is used and increase the ability of integration remanufacturing into a production of new units.

4 Conclusion

This paper describes a new model for product reuse, where parts are remanufactured and used in the production of new products. It addresses the challenge of customer perception regarding remanufactured products. Several researchers have examined the willingness to pay for remanufactured products. In all studies, the achieved price was less than for new products [18-21]. Customers perceive remanufactured products as used ones entailing lower benefits and higher risks [21, 22]. The paper describes the opportunity to overcome this challenge by providing less information and addressing the visual appeal in the treatment process. The approach is explained based on the example of a washing machine. The bought product is always a new one, as it consists of approved assemblies and goes through the standard production process. It also guarantees the predetermined functionalities, appearance and warranty. If no further information is provided, customers do not experience any further difference. This idea can be compared to the use of recycled materials in production. In general, customers cannot determine the proportion of recycled materials. Only in exceptional cases, such as in the case of children's toys, is information provided stating that the product consists only of new materials.

To implement this model, a consideration of the legal situation is necessary. There is no legal definition of used products. For example, the Higher Regional Court of Hamm (Oberlandesgericht Hamm) has decided that a product is considered a used product (Gebrauchtware), if it has been used under normal conditions and is therefore associated with a higher risk of defects [29]. Used products are allowed to have a reduced warranty, which influences the customer perception. Remanufactured products have been used under normal conditions, but have no higher risk of defects than new products and come with at least the same warranty as a new one. However, they still cannot be described as new products. In this context, a Dutch court has ruled that remanufactured products cannot be used as exchange products as part of the warranty, because they are not an equivalent substitute for a new product [30].

5 Zusammenfassung

Der Mensch verbraucht bereits jetzt mehr Ressourcen, als die Erde regenerieren kann. Zusätzlich wird der Bedarf an Rohstoffen weiter steigen. Die Weltwirtschaft folgt hauptsächlich den linearen Regeln, bei denen nur ein Lebenszyklus eines Produktes berücksichtigt wird. Remanufacturing, die Wiederverwendung von Produkten, mit einem Umsatz von rund 30 Milliarden Euro in Europa macht lediglich einen kleinen Teil des Umsatzes der gesamten Produktion aus. In diesem Beitrag wird die Integration von Remanufacturing in die Produktion von neuen Gütern vorgestellt, um eine Verlagerung von wiederaufbereiteten Produkten von Gebrauchtware zu Neuware zu ermöglichen. Die Chancen dieses Ansatzes werden am Beispiel von Haushaltsgeräten erläutert.

6 References

[1] Pkw-Bestand in weltweiten ländern in den Jahren 2005 bis 2015 (in 1.000 Pkw); https://de.statista.com/statistik/daten/studie/415061/umfrage/pkw-bestand-in-weltweiten-laendern/ ; last access: 18.05.18

[2] Business Wire; Absatz von Smartphones weltweit in den Jahren 2009 bis 2017 (in Millionen Stück). https://de.statista.com/statistik/daten/studie/173049/umfrage/weltweiter-absatz-von-smartphones-seit-2009/ last access: 18.05.18

[3] Statistisches Bundesamt; Umwelt-Abfallbilanz 2015 (Abfallaufkommen/-verbleib, Abfallintensität, Abfallaufkommen nach Wirtschaftszweigen); 2017

[4] de Wit, M. et al.; The Circularity Gap report. An analysis of the circular state of the global economy; January 2018

[5] Umweltbundesamt; Abfallaufkommen; https://www.umweltbundesamt.de/daten/ressourcen-abfall/abfallaufkommen#textpart-2; last access 14.03.2018

[6] United Nations; World population projectes to reach 9.8 billion in 2050, and 11.2 billion in 2100; published: 21.06.2017; https://www.un.org/development/desa/en/news/population/world-population-prospects-2017.html; last access 24.04.2017

[7] Nijland, H. et al.; Mobility and environmental impacts of car sharing in the Netherlands; In:Environmental Innovation and Societal Transitions Vol. 23, p. 84-91, 2017

[8] Steinhilper, R.; Remanufacturing: The Ultimate Form of Recycling. Stuttgart, Germany: Fraunhofer IRB Verlag, 1998.

[9] Gutowski, T. et al.; Remanufacturing and Energy Savings; In: Environmental Science and Technology; 45, p. 4540-4547, 2011

[10] Lund, R.T.; Remanufacturing: The Experience of the United States and Implications for Developing Countries. World Bank Technical Paper, Vol.31, 1985

[11] Sundin, E.:Product and Process Design for Successful remanufacturing. Dissertation. Departmnet of Mechanical Engineering, Linköping University; 2005

[12] Nasr, N. et al.; remanufacturing: a Key Enabler to Sustainable Product System; In:Proceedings of the 13th CIRP International Conference on life Cycle Engineering (LCE), p.15-18, 2006

[13] Remanufactured Goods: An Overview of the U.S. and Global Industries, Markets and Trade; Investigation number 332-525, USITC Publication 4356, United States International Commission; 2012

[14] Parker, D. et al; Remanufacturing Market Study; November 2015

[15] Lange, U.; Ressourceneffizienz durch Remanufacturing - Industrielle Aufarbeitung von Altteilen; Kurzanalyse Nr. 18, 2017

[16] Widera, H.; Geschäftsmodelle der Wiederaufbereitung für Hersteller von Originalteilen; Berichte aus dem Produktiontechnischen Zentrum Berlin; 2014

[17] Tukker, A.; Eight Types of Product-Service System: eight way to sustainability? Experiences from SUSPRONET; IN: Business Strategy and the Environment; Vol. 13, p. 246-260; 2004

[18] Agrawal, V., et al.; Remanufacturing, third-Party Competition, and Consumers‘ Perceived Value of New Products; In: Management Science 61(1):60-72. https://doi.org/10.1287/mnsc.2014.2099; 2015
[19] Rathore, P., et al.; Sustainability through remanufacturing in India: a case study on mobile handsets; In: Journal of Cleaner Production 19, p. 1709-1722, 2011
[20] van Weelden, E., et al.; Paving the way towards circular consumption: exploring consumer acceptance of refurbished mobile phones in the Dutch market; In: Journal of Cleaner Production 113, p.743-754, 2016
[21] Matsumuto, M. et al; Comparison of U.S. and Japanese Consumers' Perceptions of Remanufactured Auto Parts; In: Journal of Industrial Ecology; Valume 21 Number 4; 2016
[22] Ackerman, D., et al.; Assuring me that it is as 'Good as New' just makes me think about how someone else used it. Examining consumer reaction toward marketer provided information about secondhand goods; In: Journal of Consumer Behaviour, J. Consumer Behav., 16: 233–241 (2017)
[23] Subramanian, R.,et al.; Key Factors in the Market for Remanufactured Products; In: Manufacturing & Service Operations Management 14(2):315-326. https://doi.org/10.1287/msom.1110.0368, 2012
[24] Stamminger, R.; et al.; Waschen in Deutschland- Auswertung einer Verbraucherbefragung; In: SÖFW-Journal, 131, 11-2015, 2015
[25] https://www.waschmaschinen-test.eu/wie-hoch-ist-die-lebenserwartung-einer-waschmaschine; last access: 20.04.2018
[26] Fachgemeinschaft für effiziente Energieanwendung e.V.; http://service.hea.de/fachwissen/waschmaschinen/1-marktdaten.php; last access: 15.04.2018
[27] General Statistics Office of Viet Nam; Data results of the Viet Nam Household Living Standards Survey 2012
[28] Handelsblatt; Tabelle: Kaufkraft der Lohnminute; http://www.handelsblatt.com/politik/konjunktur/butter-zucker-tageszeitung-tabelle-kaufkraft-der-lohnminute/2937670.html; published 24.03.2008; last access 14.03.2018
[29] OLG Hamm, verdict from 16.01.2014 - 4 U 102/13; https://medien-internet-und-recht.de/volltext.php?mir_dok_id=2567
[30] reman IPad cannot settle court dispute; https://www.rematec.com/news/news-articles/reman-ipad-cannot-settle-court-dispute/; published 22.05.2017; last access: 27.04.2018

Planned Obsolescence in Portable Computers
- Empirical Research Results -

Martin Adrion[1] and Jörg Woidasky[1]
[1]School of Engineering, Pforzheim University, Pforzheim, 75175, Germany
joerg.woidasky@hs-pforzheim.de

Abstract

The goal of the study presented here was to use empirical research to verify and to quantify planned obsolescence effects, and specifically to identify the role of user expectations. The products subject to this study were chosen to be portable devices ("notebooks") such as laptops, notebooks, netbooks, and ultrabooks, which were chosen because they are directly used by consumers in large scale, and exhibit a sufficient technical complexity. The population researched were students of a university in southwestern Germany. For data acquisition, a questionnaire with 29 single questions was drafted and pre-tested in a student environment. It took about 15 minutes for filling in.

Indicators measured covered amongst others the life span expectations of the computer, the life span of the previous (not the actual) computer model used, the time of repairs, the warranty period length, and the reason for replacing the last computer.

The interviewees were selected from the Bachelor and Master students of the university applying three-stage cluster sampling. The questionnaire was handed out to a total of 215 students, with a return rate of 99%. Out of the sample of n=212 students 59% stated that they had owned another notebook model before the actual one. In case they had pre-owned a notebook computer, their reasons for replacing it was a technical fault (49%), the fact that it was outdated (40%), missing compatibility (6%) and to 5% other reasons (n=126).

In the researched population, notebook use phase spans range from 3.3 years to 4.6 years. The arithmetic mean use phase length was found to be 4.0 years. Students which read the directions for use experienced a longer notebook use phase (4.6 years). Those students who undertook maintenance operations on their notebook experienced shorter use phase spans (only 3.7 years). With regard to repairs, mainly battery replacement, display repair and the power supply, software, and keyboard repairs were of major importance. The second use phase year was found to be the most repair-intensive phase of the notebooks, although about 65% of all notebooks worked flawlessly without any repair.

Comparison of expectations and observations of use phase lengths yielded the result that 64% of all notebook users expect a longer use phase than they had actually experienced with their previous product. The difference between expectation and observation is highest with the users of Fujitsu Siemens, Apple, Lenovo and Asus notebooks. Brands with a long use phase length seem to be Medion and possibly Toshiba.

As a result, about two thirds of the notebook users might claim to be subject to planned obsolescence due to a mismatch of expected and delivered use phase length, although from a merely technical point of view a sufficient notebook performance might have been delivered which clearly exceeded the warranty period.

1 Introduction

Obsolescence of electrical and electronical products has become a research topic in recent years [1]. Mainly in its variant "planned obsolescence" it has become common understanding to accept the assumption that products are being built intentionally faulty or below their technical potential regarding durability. Proof for this knowledge is scarce, as most of the information is based on individual observations and anecdotes [2].

Intentionally lowering the life span of products, i. e. "planned obsolescence" is said to be favorable for any producing company, as a short product life span leads to a demand increase by customers and/or users over time. Moreover, product quality and life span are correlated to a certain extent to the production and component costs, and thus the consumers' willingness to pay may indirectly influence product quality as well. One aspect though which little attention was paid to recently, is the role of the user expectations when the discussion on planned obsolescence arises.

From a scientific point of view, the technical life span and the use phase length have to be distinguished: The technical life span covers the period of time in which the product is operational and ready to use (i. e. the time between initial purchase and defect of the product). In contrast, the use phase length is not governed by technology, but by the user deciding to discontinue the products use, which may happen at the end of the technical life span, or at an earlier point in time, while the product may still be operational.

Obsolescence of products is a "natural" effect, although it is observed in a technosphere environment. Degradation of materials or wear due to mechanical stresses or other reasons occur with any product. Both the technical product properties designed and built by the producer and the consumers use patterns are relevant for the technical life span of any product. As a consequence, research on planned obsolescence has to observe both technical aspects of the life span limitation and the users behavior, as obviously any product is designed to be used within defined operational limits such as environmenal conditions (e. g. temperature, humidity), servicing (e. g. cleaning, greasing) or maximum stress levels.

A. Pehlken et al. (Eds.), *Cascade Use in Technologies 2018*,
https://doi.org/10.1007/978-3-662-57886-5_3

A systematic of obsolescence was given by Packard [3] as early as the 1960s, identifying the functional, the qualitative and the psychological obsolescence. Functional obsolescence is observed when an operational product becomes obsolete due to technical innovations (such as a typewriter becoming obsolete due to computer use). Qualitative obsolescence is observed when a product failure occurs which renders the product obsolete (such as a light bulb becoming obsolete due to wire fusing). Psychological obsolescence is observed when the user ends the product use due to personal preferences although the product might still be operational.

Thus an effort is undertaken to empirically try to confirm the common understanding that "planned obsolescence of electrical products doubtlessly exists". Hence the goal of the study presented here was to use empirical research to verify and to quantify planned obsolescence effects with portable computers, and especially to research the role of consumer expectations. Portable devices cover laptops, notebooks, netbooks, and ultrabooks, which were all subject of this study. The characteristic elements of these products were integrated display, keyboard, touchpad and battery, and an external power supply.

Portable computers ("Notebooks" in the following) were chosen as research objects as they are used in a business-to-consumer environment and thus may be researched by e. g. interviewing a students' population. Moreover, they are electronical products with a high complexity and comparatively high number of interacting electrical components, which should enable an easy identification of planned obsolescence effects.

2 Methods

The theoretical construct [4] to describe is the planned obsolescence of electrical and electronical products, using mobile computers as an example. The dimensions to be researched cover the consumers' expectations, the actual life span, repairs of the computers, warranty aspects, and actual observations of obsolescence. These dimensions formed the basis for the indicators to be measured, which cover the life span expectations of the computer users, the life span of the last (not the current) computer model used, the time of repairs, the warranty period length, and the reason for discarding the last computer. The core question to be answered was if results can be found which support the hypothesis that planned obsolescence occurs in mobile computers.

2.1 Questionnaire Design

The questionnaire was divided into five sections, covering questions (a) on obsolescence in general, (b) on notebook computers in general, (c) on the present notebook, (d) on the previously used notebook, if any, and (e) on personal data. After a short informative written statement, one easy-to-answer introductory question was asked, with specificity and complexity of the subsequent questions increasing. A second text statement expressing thanks for participation finalized the questionnaire which featured 29 questions in total. A copy of the questionnaire may be obtained from the author. Selected questions are presented in the results section of this paper.

A pretest on this questionnaire was carried out by randomly selecting 15 students both male and female of the university campus, and having them fill in the questionnaire. Besides minor revisions, this text yielded an individual duration of about 15 minutes for filling in the questionnaire.

2.2 Interviewees Selection

The interviewees were selected from the Bachelor and Master students of a university in southwestern Germany, as it can be expected that a high share of this population owns and operates portable computers. Moreover, access to this population was easily available, and a good compliance and high answering rate could be expected. Three-stage cluster sampling was applied, by first randomly choosing six master and eight bachelor programs from of a total of 12 master and 21 bachelor programs from two schools of the university. The second step covered the random selection of semesters, and the third step the selection of specific classes. The lecturers of these classes have been approached upfront, requesting access to the students during their classes for about 15 min to fill in the questionnaires. All lecturers cooperated.

A questionnaire procedure was chosen for data acquisition. The questionnaires were handed over to the interviewees in class. While filling in, a contact person was available in the room to answer questions on the spot. The filled in questionnaires were collected, coded and transferred into an SPSS file, using the IBM SPSS Statistics 21 software. The entire work was carried out in May and June 2015 in German language.

3 Results and Discussion

A total of 215 students answered the questionnaires, with a return rate of 99%, as only three students did not return their questionnaires. As not all questions had been answered in all questionnaires, the n figure might be lower than 212. The share of female students answering was 51% (n=210). To describe the sample in more detail, the students were asked for a self-assessment regarding their social milieu, using the SINUS milieus [5] (Table 1). The students were introduced to the milieu definition by characterization of each milieu with a keyword and a two-line characterization. Moreover, a comprehensive graph was presented along with the question, displaying the milieu allocation in a portfolio of social stratum (lower/middle/uppler class) versus attitudes towards innovations (tradition/modernization and individualization/reorientation).

Table 1. Students self-assessment on SINUS milieu affiliation
(n=197; Question "Please try to allocate yourself to one of these milieus", after a short milieu explanation)

SINUS milieu affiliation	Students figure (absolute numbers)	Students share (%)
Established	23	11.7
Liberal Intellectuals	31	15.7
Performers	19	9.6
Cosmopolitan Avantgarde	14	7.1
Modern Mainstream	38	19.3
Adaptive Navigators	43	21.8
Social-Ecologicals	8	4.1
Traditionals	6	3.0
Precarious	1	0.5
Hedonists	14	7.1
TOTAL	197	100.0

The arithmetic mean life span of the last notebook computer of female users was 4.2 years, while the male users' computers were used for 4.0 years. This difference was proven to be not significant (due to a biserial correlation test with p of 0.574).

Out of the sample of n=212 students 59% stated that they had owned another notebook model before the current one. In case they had pre-owned a notebook computer, their reasons for replacing it was a technical fault (49%), the fact that it was outdated (40%), missing compatibility (6%) and to 5% other reasons (n=126). The following sub-chapters will present additional selected results.

When being asked "How often do corporations take deliberate steps to move consumers to replace an old device by a new one?", 32% of the interviewees (n=212) expected this to happen "very often", 51% "often", 13% "sometimes", 4% "seldom" and 0,5% "never".

3.1 Notebook Use Patterns

The individual use pattern of the computer use can be derived from Table 2. About one third of all students used their computers for longer than four hours a day.

Table 2. Students laptop computer use pattern
(n=207; Question "How many hours per day is the current device being used?")

Daily Laptop use	<1 h	1-2 h	2-3 h	3-4 h	>4 h
Students share (%)	13	16	22	17	32

Table 3. Relevance of reading the directions for use and use according to specifications for product use phase length
(Cross-classified table using questions on the present notebook "Did you read the directions for use before operating your device?" and "Do you feel you are using the device according to its specifications?" along arithmetic mean use phase length of previous notebook given in brackets)

absolute figures	Using the device according to specifications (in brackets: mean use phase length of previous notebook in years)			
Read directions for use	Yes	No	do not know specifications	TOTAL
Yes	27 (4.6a)	3 (4.2a)	7 (3.8a)	37
No	105 (4.0a)	23 (4.2a)	39 (3.3a)	167
TOTAL	132	26	46	204

Designated use of products is a key requirement for a long product use span. Consequently, the interviewees were asked if they had read the directions for use of their computer. Moreover, in another question they were asked if they thought that they used their computer according to its specifications. The combined results from both questions can be found in Table 3. Moreover, the arithmetic mean values of the use phase span of the last notebook in years of the interviewees are given in brackets. These use phase spans range from 3.3 years to 4.6 years and show clearly a correlation between knowledge and compliance regarding intended notebook use. From all users who feel that they are using the devices according to the specifications, the notebooks of users who read the instructions are experiencing a 0.6 years longer use phase span (4.6 versus 4.0 years).

3.2 Maintenance Activities

The question "Were there maintenance activities carried out during the use phase of the device?" was answered by 86% of the interviewees by "no", a mere 14% answered with yes (n=206). From those who answered yes, the individual actions taken were in absolute figures, as multiple answers to this question were possible, software updates (mentioned 10 times), virus removal (8), fan cleaning (7), working memory extension (4), hard disc drive replacement (3). Graphic board driver update and hard disc drive defragmentation each have only been mentioned once, amounting to a total of 34 maintenance activities in the sample. Regarding the mean use phase length of the previous notebook, surprisingly the computers of those users carrying out maintenance activities on their actual notebook were used only 3.7 years, whereas the computers of those who did not carry out any maintenance reached 4.0 years.

3.3 Warranties

For the previous notebook model, the warranty period duration was asked. The answers ranged between one year (17%), two years (54%) and five years (1%), with 28% of the interviewees stating "unknown" (n=126). Arithmetically a mean warranty period of 1.8 can be calculated, which contrasts clearly to the arithmetic mean of the use phase length of the previous notebook of 4.0 years. As in Germany a two year period of warranty is legally required, this may be taken as a quasi standard value, which may be extended upon the producers or retailers choice, but not cut short. Comparing these warranty period lengths it can be stated that the arithmetic mean value of the use phase length (4.0 years) exceeds the warranty period by far.

3.4 Repairs

Regarding repair frequency, both the values for the actual and the previous notebook are on display in Table 4, along with the repair distribution over time. It was found that 65% of all (previous) notebooks never underwent a repair. Out of those repaired, the highest figure of repairs occurred in the second and third year after purchase. As the rechargeable battery is one of the components highly prone to repair and replacement, its time of replacement was studied. Out of the eleven batteries replaced in the previous notebooks, only one each was replaced in the first, third and fifth year of the use phase, whereas in the second year six and in the fourth year two batteries were replaced. This observation might be motivated by replacement activities triggered by the warranty period of two years nearing its end. The same might be the case for the maximum of 15 single repair activities observed in the second year after purchase of the previous notebook (Table 4 part II).

Table 4. Repair frequencies and time distribution of present and previous notebooks

		present notebook	previous notebook
I-How often was/is the device repaired? (relative figures)		*(n=207)*	*(n=122)*
	Never	78%	65%
	Once	19%	20%
	More than once	3%	15%
II-How many years after purchase are the devices repaired? (absolute figures)			
	< 1a	16	6
	1-2a	8	15
	2-3a	9	11
	3-4a	8	8
	>4a	5	4

Figure 1 gives the results of an open question on the repair actions of the present and the previous notebooks. Regarding the previous notebook whose use span has already ended, the battery replacement ranks first, the display repair second and the power supply, software, and keyboard all rank third in importance.

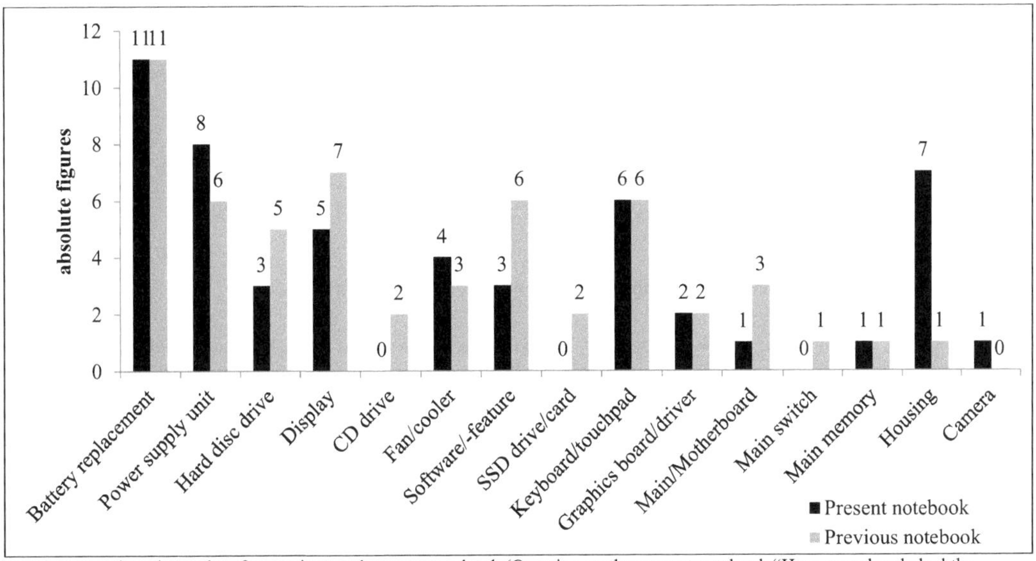

Figure 1. Repair actions taken for previous and present notebook (Question on the present notebook "Have you already had the device or its components (incl. battery, power supply, cord) repaired?" and on the previous notebook "What exactly did you have repaired?")

3.5 Use Phase Length Expectations and Observations

User dissatisfaction may originate from products which offer a service life shorter than expected. Consequently in the study not only the devices' actual use phase length, but also the users' expectations regarding the use phase length have been identified. The arithmetic mean value of expected use phase length of all interviewees including those without previous notebooks was at 5.3 years (n=211), whereas the actual use phase of all previously used notebooks was found to be 4.0 years (n=124). Figure 2 shows the combined data set, with individual expectation-observation data twins plotted in the graph. Out of the total figure of 124 twin data sets, in merely 20 cases (16%) the expectation is identical with the use phase length observation, thus coinciding with the bold black line in Figure 2. In the graph's portion below this line 25 data twins (20%) are allocated, representing users with devices exceeding the users use phase lengths expectation. Consequently, 79 users (64%) have experienced a shorter use phase than expected from their notebook. Regardless of the justification of the expectations, it can be stated that these almost two thirds of the notebook users (64%) form the potential for planned obsolescence arguments.

Eventually from a user and consumer perspective, a brand specific view might be of relevance. To this end, questions on the expected and actual life span of the previous notebook have been combined with the information on the producer of this previous product. The brands covered were HP (n=17), Apple (7), Dell (10), Samsung (16), Lenovo (7), Asus (20), Acer (19), Sony (7), Medion (7), Fujitsu Siemens (5), Toshiba (2) and others (4). As the total figure of datasets of some of the brands is limited, such may be the reliability of these brand specific arithmetic mean values in Figure 3. Nonetheless the comparison of the arithmetic mean value of actual use phase lengths of the individual brands given in grey bars in Figure 3 with the diamond shaped marks for the equivalent expected use phase length reveals that all but one brand fail to meet or to exceed the users expectations. Only in the case of Toshiba is this not the case, but due to only two sets of data for this brand statistical proof is missing. With a mean use phase length of 5.2 years the Medion products are near the arithmetic mean value of expected use phase length of all interviewees, which was found to be 5.3 years. The Fujitsu Siemens users exhibit the highest expectations (6.0 years), and the highest difference of expected and observed use phase (1.7 years), followed by 1.6 years difference values for Apple, Lenovo, and Asus. The lowest differences can be found with 0.3 years for Sony products and 0.2 years for other products.

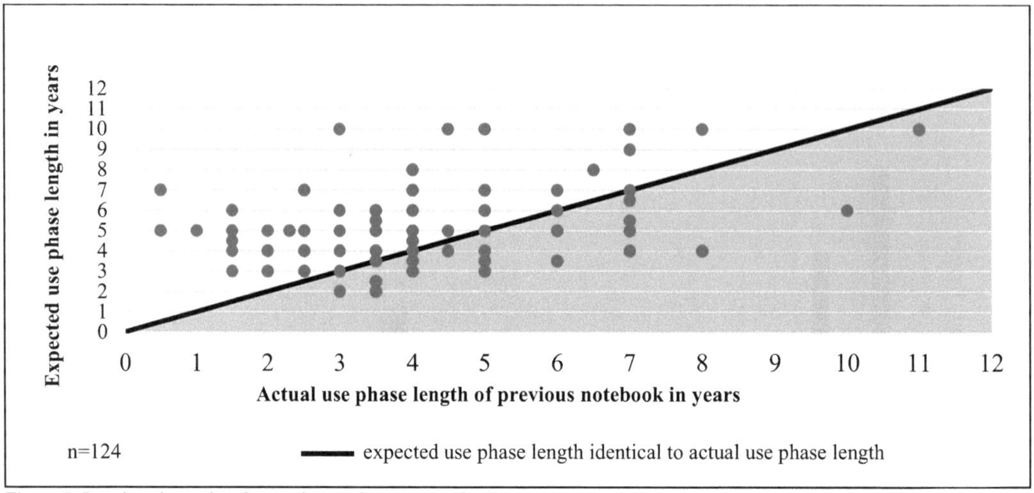

Figure 2. Repair actions taken for previous and present notebook
(Combination of question on notebooks in general "How many years of use phase length do you expect from a notebook?" and on the previous notebook "How many years did you use the device?")

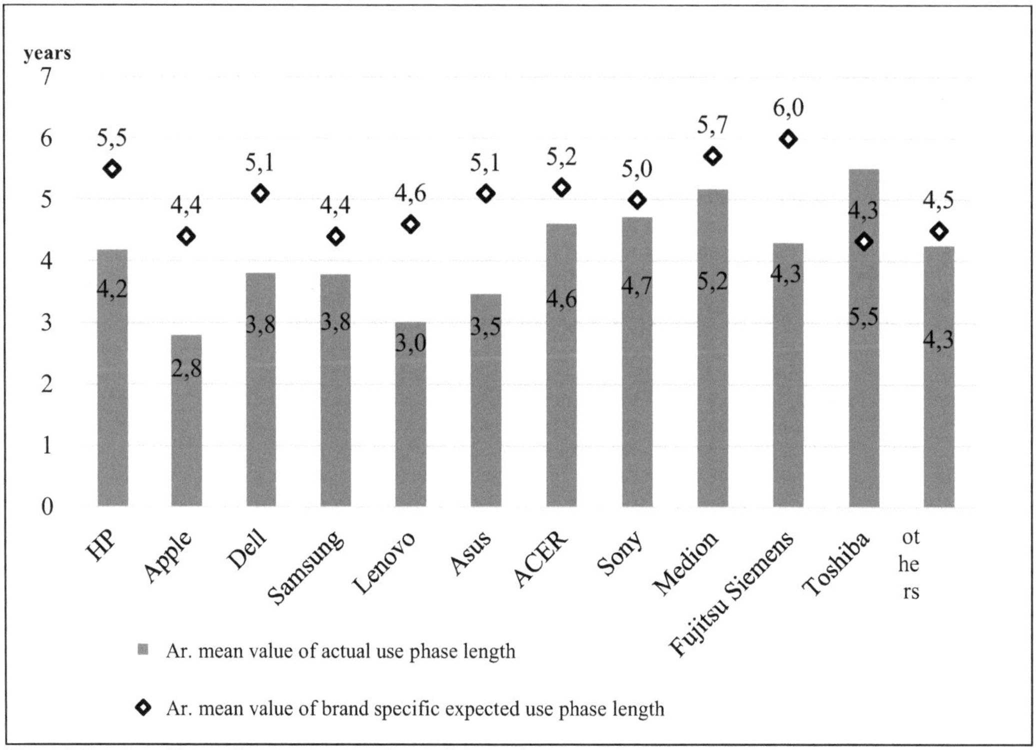

Figure 3. Brand specific use phase length expectations and observations
(Combination of questions on notebooks in general "How many years of use phase length do you expect from a notebook?" and on the previous notebook "How many years did you use the device?" along with "Who was the producer of the previous product?")

4 Conclusions

Although the survey results fail to deliver clear proof of planned obsolescence in the case of notebook computers, some remarkable facts could be extracted from the study:
In a university student environment, notebook use phase spans range from 3.3 years to 4.6 years. The arithmetic mean use phase length was found to be 4.0 years. Students which read the directions for use experienced a longer notebook use phase (4.6 years). Interestingly those students who undertook maintenance operations on their notebook experienced shorter use phase spans (only 3.7 years). With regard to repairs, mainly battery replacement, display repair and the power supply, software, and keyboard repairs were of major importance. The second use phase year seems to be the most repair-intensive phase of the notebooks, although about 65% of all notebooks worked flawlessly without any repair.
Comparison of expectations and observations of use phase lengths yielded the result that 64% of all notebook users expect a longer use phase than they actually have experienced with their previous product. The difference between expectation and observation is highest with the users of Fujitsu Siemens, Apple, Lenovo and Asus notebooks. Brands with a long use phase length seem to be Medion and possibly Toshiba.
As a result, about two thirds of the notebook users might claim to be subject to planned obsolescence due to a mismatch of expected and delivered use phase length, although from a merely technical point of view a sufficient notebook performance might have been delivered which clearly exceeded the warranty period. In fact, the survey showed that more than 80 % of all interviewees expect planned obsolescence to be applied by laptop computer producers often or very often.

5 Zusammenfassung

Das Ziel der hier vorgestellten Untersuchung war es, das Auftreten geplanter Obsoleszenz nachzuweisen und mit Hilfe empirischer Methoden zu quantifizieren, und dabei insbesondere auf die Rolle der Nutzererwartungen einzugehen. Als Untersuchungsgegenstände wurden tragbare Computer („Notebooks“) wie Laptops, Notebooks, Netbooks und Ultrabooks gewählt, da sie von Verbrauchern in großem Umfang direkt genutzt werden und eine hinreichende technische Komplexität aufweisen.
Die untersuchte Grundgesamtheit stellte die Studierendenschaft einer Hochschule in Südwestdeutschland dar. Zur Datenerhebung wurde ein 29 Fragen umfassender Fragebogen erstellt und einem Pretest unterzogen. Das Ausfüllen des Fragebogens benötigte etwa 15 Minuten.
Die so untersuchten Indikatoren umfassten unter anderem die erwartete Lebensdauer des mobilen Computers, die Nutzungsdauer des vorherigen und aktuell genutzten Computermodells, Reparaturzeitpunkte, die Garantiedauer sowie die Gründe für den Austausch des vorherigen Computers.
Die Befragten wurden aus Bachelor- und Masterstudiengängen mit Hilfe einer dreistufigen Klumpenauswahl zufällig ausgewählt. Insgesamt wurde der Fragebogen an 215 Studierende ausgegeben. Die Rückgaberate lag bei 99%. Von den antwortenden Studierenden (n=212) gaben 59% an, sie hätten vor dem jetzigen bereits ein anderes Notebook besessen. Die Gründe für den Austausch des vorherigen Modells waren dessen Defekt (49%), die Einschätzung, das Gerät sei nicht mehr zeitgemäß (40%), dessen fehlende Kompatibilität (6%) und zu 5% andere Gründe (n=126).
In der befragten Gruppe wurden Nutzungsdauern zwischen 3,3 und 4,6 Jahren angetroffen. Das arithmetische Mittel der Nutzungsdauern lag bei 4,0 Jahren. Studierende, die die Gebrauchsanweisung des Gerätes lasen, konnten ihre Geräte länger nutzen (4,6 Jahre). Hingegen lag die Nutzungsdauer der Studierenden, die Wartungsprozesse an ihrem Gerät durchführten, mit 3,7 Jahren deutlich darunter. Die relevantesten Gerätereparaturen sind der Austausch von Batterien, die Reparatur des Displays und von Netzteilen und Software sowie von Tastaturen und Touchpads. Die meisten Reparaturen wurden im zweiten Nutzungsjahr der Geräte durchgeführt. Insgesamt arbeiteten jedoch 65% aller Notebooks störungsfrei ohne jegliche Reparatur.
Der Vergleich der erwarteten und der tatsächlichen Nutzungsdauer zeigte, dass 64% aller Nutzer von Notebooks eine längere Nutzungsdauer erwarten als jene, die von ihrem letzten Gerät tatsächlich erreicht wurde. Die größten zahlenmäßigen Differenzen von erwarteter und erreichter Lebensdauer wurden bei Nutzern der Marken Fujitsu Siemens, Apple, Lenovo und Asus festgestellt. Marken wie Medion und möglicherweise Toshiba scheinen eine vergleichsweise lange Nutzungsdauer aufzuweisen.
Im Ergebnis kann festgestellt werden, dass nach den Befragungsergebnissen etwa zwei Drittel der Notebook-Nutzer die Behauptung aufstellen könnten, von „geplanter Obsoleszenz“ betroffen zu sein, da ihre Erwartungen an die Nutzungsdauer ihres Notebooks nicht erfüllt wurde, obwohl aus technischer Sicht kein Mangel vorliegt, was z. B. anhand einer deutlichen Überschreitung des Garantiezeitraums deutlich wird.

6 References

[1] Prakash, Siddharth; Dehoust, Günther; Gsell, Martin; Schleicher, Tobias; Stamminger, Rainer: Einfluss der Nutzungsdauer von Produkten auf ihre Umweltwirkung: Schaffung einer Informationsgrundlage und Entwicklung von Strategien gegen „Obsoleszenz“. UBA-Texte 11/2016. Dessau-Roßlau, Februar 2016

[2] Schridde, Stefan; Kreiß, Christian; Winzer, Janis: Geplante Obsoleszenz: Entstehungsursachen, Konkrete Beispiele, Schadensfolgen, Handlungsprogramm. ARGE REGIO Stadt- und Regionalentwickung GmbH 2013. Gutachten im Auftrag der Bundesfraktion Bündnis 90/Die Grünen, S. 28-49.
[3] Packard, Vance: Die große Verschwendung. Econ Verlag. Düsseldorf, 1966
[4] Jacob, Rüdiger; Heinz, Andreas; Décieux, Jean Phillipe; Eirmbter, Willy H.: Umfrage – Einführung in die Methoden der Umfrageforschung. Oldenbourg Wissenschaftsverlag GmbH, 2. Auflage. München, 2011
[5] https://www.sinus-institut.de/en/sinus-solutions/sinus-milieus/ as of March 8, 2018

Gaps and Needs within the WEEE Management in Brazil

Sascha Diedler[1] and Kerstin Kuchta[1]
[1]Institute of Environmental Technology and Energy Economics, Hamburg University of Technology, Hamburg, 21079, Germany
sascha.diedler@tuhh.de

Abstract

The generation and processing of waste electrical and electronic equipment (WEEE) has given rise to numerous challenges around the world. Accordingly, preventive approaches basen on governmental and environmental regulatory pressure are highly relevant to treat WEEE in a proper manner. Within Latin America (LATAM) Brazil produces the highest total amount of WEEE with a still rising tendency, while recycling schemes are still in their infancy [1]. Therefore, it is important to outline the main gaps and needs of related systems. Within the framework of this paper, recommendations for improving the management of WEEE in Brazil are derived based on the expertise of various local experts in this sector. The results of the knowledge sharing methodology of a World Café show that political enforcement of implemented national laws and close cooperation of different stakeholder groups are key instruments to improve the general WEEE management in Brazil.

1 Introduction and Theoretical Background

17 Sustainable Development Goals (SDGs) were implemented in 2015 by the United Nations (UN) as follow ups of the millennium goals to ensure global sustainable development. Every member country of the UN is responsible for reaching the same measurable indicators of the 169 targets, which means that there is no differentiation between developed or developing countries [2]. To reach the SDGs in the frame of the agenda 2030 it is highly applicable that every member country of the UN cooperates closely by using synergies within good practices of various sectors of action effecting the global sustainable development. Waste management and its proper treatment marks one of those sectors, and has to be tackled in every society around the world. For decades, the fastest growing waste stream has been WEEE.

Cucchiella et al. [3] estimates the global WEEE generation at 20 – 50 million tons per year and claims an annual rising rate of 5%. According to Baldé et al. [4] the total amount of WEEE generated worldwide in 2014 was 41.8 million tons, of which the highest amount was generated in Asia (38.3%) followed by America (28.0%) and Europe (27.7%). The quantity in LATAM is estimated at 3.9 million tons, which indicates an amount of 6.6 kg per capita [1]. As WEEE may contain hazardous components [5] as well as valuable resources, its treatment and disposal may cause emerging environmental challenges but also business opportunities.

Because of this, the quality of WEEE management respectively treatment affects various SDGs directly, so that Goal 3 (good health and well-being) and Goal 6 (clean water and sanitation) are touched directly. Sustainable cities and communities (Goal 11) is affected due to its target of reducing the environmental impact per capita through appropriate treatment of municipal and other waste streams in cities. The target of responsible consumption and production (Goal 12) aims at an environmentally sound management of waste, especially chemicals, which may cause negative impacts on human health and the environment. In addition to that, special emphasis is put on reducing general waste amounts through prevention, reuse or at least recycling practices. As WEEE contain valuable resources like copper, silver or gold, WEEE management schemes mark business opportunities [6]. This is why goal 8 (decent work and economic growth) is directly affected by WEEE and its treatment as well. [7]

1.1 WEEE Situation in LATAM

Because of the import restrictions imposed by China on 24 types of solid waste in 2018, waste companies around the world are searching for new pathways to dispose of their waste. As the restrictions will also affect scrap metal imports, LATAM is considered as a new frontier in terms of material processing according to Acosta [8], as it mirrors structures and the potential of growth of China 20 years ago. LATAM is characterized by a high urbanization rate which causes a high market penetration rate of IT and other electronic equipment [1]. This results in a rapid growth of sales of electrical and electronic equipment (EEE), which in turn leads to increasing quantities of WEEE [4].

Even if LATAM generates less WEEE than other continents, it is of high interest to follow up on its development, as legislation and policy developed rapidly recently. In fact, every country in LATAM has ratified the Basel convention, so that all countries tackle WEEE-related topics within their national policy. Until now, specific WEEE legislation is implemented in Brazil, Colombia, Costa Rica, Ecuador, Mexico and Peru, as shown in Figure 1. The figure further indicates that Argentina, Bolivia, Chile, the Dominican Republic, El Salvador, Guatemala, Honduras, Nicaragua, Panama, Uruguay and Venezuela have a specific WEEE legislation in development. In the frame of the SDGs and the global interest in processing WEEE properly and further circulating its valuable resources, it is reasonable to use

A. Pehlken et al. (Eds.), *Cascade Use in Technologies 2018*,
https://doi.org/10.1007/978-3-662-57886-5_4

synergies and find good practices of WEEE processing and implement analoguous legislation and practices in the area of LATAM.

Figure 1. Status of ratification of the Basel convention and WEEE legislation [1]

Based on estimations of Magalini et al. [1], the total amount of WEEE in LATAM amounts to 3904 kt, of which the highest amounts are categorised to small and large household equipment. Comparing the WEEE amount per category with the global amount, which amounts to a total of 41800 kt in 2014, indicates that the composition of WEEE generated in LATAM is very much the same as WEEE that occurs worldwide, as visualized in Figure 2.

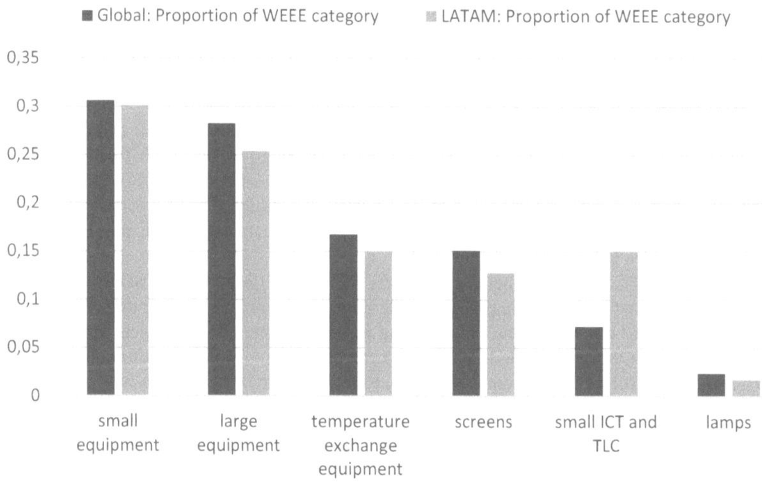

Figure 2. Proportion of total WEEE generated per category worldwide and in LATAM in 2014

Figure 2 shows the proportion of total WEEE generated per category by comparing the global amount with the amount in LATAM. As the shares are almost the same in all categories, i.e. only the proportion of small information and (tele)communication technology (ICT and TLC) equipment is much higher in LATAM, it is reasonable to derive basic structures of WEEE management systems from good practices in other countries. Therefore, it is important to expand knowledge such as the Erasmus+ funded project "Latin American-European network on waste electrical and electronic equipment research, development and analyses" (LaWEEEda) targets by aiming at improving the dialogue between researchers and practitioners in the field of WEEE management and implement European high quality training materials regarding sound WEEE management in developing countries, in their case in Brazil and Nicaragua.

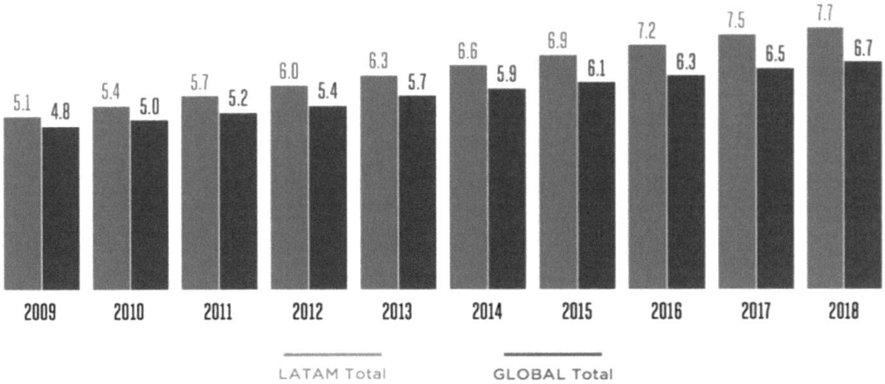

Figure 3. Comparison of the annual WEEE amount in kg per person in LATAM and worldwide [1]

Figure 3 visualizes the annual amount of WEEE in kg per capita comparing global and LATAM generation. The steady increase of WEEE per capita in LATAM is about 6% and thus slightly higher than in the rest of the world, which amounts to about 5%. Another indicator showing the constant growth within last decades of the EEE consumption resulting in increasing WEEE amounts is indicated by the development of the gross domestic product (GDP) in LATAM and the urbanization rate over the last decades shown in Table 1.

Table 1. Development of urbanization rate and GDP in LATAM

Increasing urbanization in LATAM		
Global urbanization rate	In 1950: 30 %	In 2015: 54 %
LATAM urbanization rate	In 1950: 41 %	In 2015: 86 %
Increasing prosperity in LATAM		
GDP Europe	In 1970: 0,855 BnU$D	In 2015: 16,33 BnU$D
GDP LATAM	In 1970: 0,177 BnU$D	In 2015: 5,35 BnU$D

Formal recycling of WEEE in Latin America is developing and traditional metal recycling companies have discovered the WEEE recycling market, but adequate processes are still in their infancy. Processed quantities are still modest, since neither the legal framework, nor the logistical infrastructure demands larger capacities. The countries in LATAM are, like other developing countries, characterized by rapid urbanization and economic growth (see Table 1) resulting in an increasing demand for consumer goods [9]. The contribution of informal activities is difficult to estimate as informal waste workers have no inherent reason, obligations or simply not the capabilities to keep records regarding their work. As formal performance data is usually not covering informal systems, official statistics do not reflect the complete picture of waste management in low-income countries. As WEEE contains high concentrations of valuable materials in comparison to other types of waste, the activity of the informal sector in this field is comparatively high. [1]

Main challenges that remain in the frame of WEEE management in LATAM are lacking capacities at various levels hindering an appropriate treatment of the increasing quantities of WEEE. Due to missing policies respectively its enforcement and formal collection, recycling and final treatment schemes, the informal sector is very active in the countries of LATAM. Therefore, it is of high interest to identify main gaps and needs to derive recommendations for different stakeholder groups.

1.2 WEEE Situation in Brazil

Regarding land area and population, the Federal Republic of Brazil is by far the largest country in LATAM. Consequently, it is not surprising that the highest total amount of WEEE is generated in Brazil with an amount of 1412 kt and the tendency of further increases as shown in Figure 4.

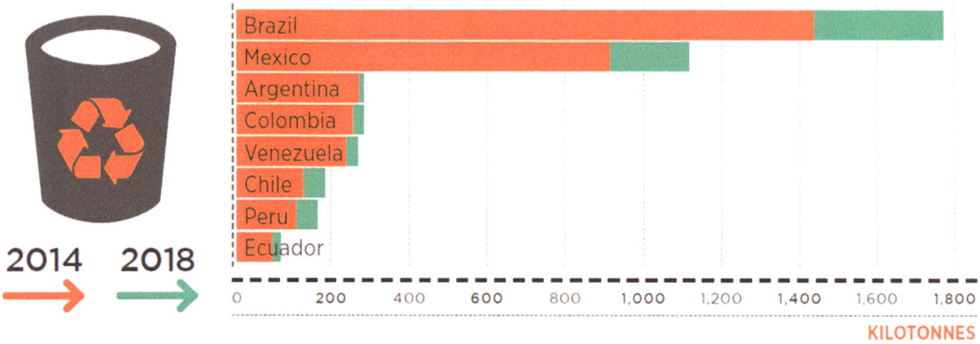

Figure 4. LATAM countries with highest total WEEE amount [1]

Regarding WEEE generated per capita, Brazilian citizens dispose about 7 kg annually. This is roughly the average amount when comparing LATAM countries, of which Chile generated the highest amount per capita with 9.9 kg annually (GSMA 2014). As shown in Figure 1, Brazil is one of six LATAM countries that have ratified a specific WEEE legislation. Within the Brazilian National Solid Waste Policy (Law No. 12305/2010), which is regulated by Decree No. 7404/2010, it is prohibited to practice open dumping and dumping on water bodies, open air or irregular burning of waste, import of hazardous waste and waste picking at waste disposal areas. In addition to that, take-back schemes for special wastes are implemented for pesticides and their waste, batteries, tires, lubricating oils and their waste, mercurial and sodium lamps and electrical and electronic equipment (EEE) and their components. WEEE take-back schemes are implemented with shared responsibilities, in which consumers are responsible for returning WEEE to retailers or distributors who will then return WEEE to manufacturers. The manufacturers in turn are responsible for an adequate WEEE destination. EEE retailers and manufacturers are further responsible for structuring and implementing take-back schemes apart from the municipal solid waste management and to keep the environmental agencies updated about their actions. [10]

Currently, 134 WEEE recycling facilities are located in Brazil. The facilities are mainly located in the Southeast of Brazil, as visualized in Figure 5. In this region, the largest Brazilian cities by population, Sao Paulo and Rio de Janeiro, are located where most of the WEEE occurs. In total, the southeastern region concentrates 58 % of recycling facilities, while the northeastern and northern regions possess 10.5 % together. Combining the northern, western and central area of Brazil, it is noticeable that in about three quarters of the Brazilian land area a maximum number of five recycling facilities is situated per region. This poses major challenges for logistics and the general infrastructure in the frame of a sustainable WEEE management and treatment scheme. [6]

Figure 5. WEEE recyclers in Brazil [6]

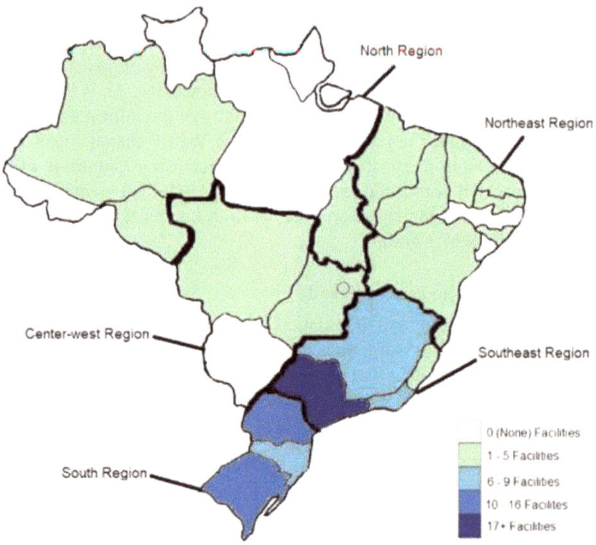

2 Methodology

As the WEEE recycling schemes in Brazil are still in their infancy, it is highly important to identify main gaps and needs from the perspective of various stakeholder groups. Within the conducted knowledge sharing method of a World Café, invited experts, covering academic, business and representatives from the official authorities of Brazil, collect and evaluate relevant aspects related to gaps and needs of the Brazilian WEEE treatment. The topics involved discussion points in terms of strengths, weaknesses, opportunities and threats related to the WEEE management in Brazil. Within this paper the results regarding the main gaps and needs in the Brazilian WEEE management are outlined. Based on the outcome, recommendations for the considered stakeholder groups are derived.

2.1 General Procedure of a World Café

The methodology of a World Café was developed by Juanita Brown and David Isaacs. This creative method is used to introduce new topics and to start constructive discussions with various people about their related approaches and ideas, which should end up with a common understanding of it [11, 12].

During a World Café, several stations are built up to discuss different aspects of a field of interest. At each station a moderator is located to introduce the relevant issue to be discussed, to make sure everybody is engaged and to collect all content that is expressed by the participants. The moderator does not differentiate between right or wrong content but notes everything that is said. After a certain time the participants switch stations while the moderator stays at their allocated station until the end of the overall World Café.

The relevance and practicability of the World Café method has been confirmed throughout various studies and is nowadays often used for strategic workshops. By bringing together experts in the field of interest and initiating the sharing of fundamental knowledge, relevant concepts are outlined tackling related challenges [12–17].

2.2 The LaWEEEda World Café

The World Café took place during the Kick-Off meeting of the Erasmus+ funded project "Latin American-European network on waste electrical and electronic equipment research, development and analyses" (LaWEEEda) in April 2017. The main idea behind this workshop was to discuss already collected information, to gain further insights, to enable first contacts between stakeholders from Brazil and Nicaragua and to engage academics, businesses and stakeholders from the administration. As LaWEEEda cooperates with two countries in LATAM it is important to mention that participants from Nicaragua took part and provided their thoughts about the Nicaraguan WEEE management at each station as well. As the data was collected separately for each country and this paper focuses on the WEEE management in Brazil, the Brazilian data will be outlined. Within the LaWEEEda World Café four stations were implemented to discuss the topics shown in Table 2.

Table 2. Topics of the four stations within the World Café

Station 1: Areas of responsibility in WEEE management			
Government	Companies	Scientists	Consumers
Station 2: Gaps and needs in WEEE management			
Government	Companies	Scientists	Consumers
Station 3: Strengths and opportunities in WEEE management			
Government	Companies	Scientists	Consumers
Station 4: Potentials to improve WEEE management within the educational sphere			
Pre-school	School	University	CPD

Station 1 aims at identifying how firm the participants are with the areas of responsibility in the Brazilian WEEE management. As pointed out in section 1.2, Brazil implemented shared responsibilities regarding the treatment of WEEE. Additionally, the participants had the chance to bring up suggestions about allocating stakeholder responsibilities. Station 2 seeks to identify the main gaps and needs regarding the Brazilian WEEE management. Like station 1, participants should allocate their thoughts and concepts to different stakeholder groups, in this case the government, companies, scientists and consumers. Station 3 was implemented to find out about the strengths and opportunities in the Brazilian WEEE management according to the participants. At station 2, the participants had to indicate their ideas to increase the quality of WEEE management in Brazil. Station 4 tackles the question about general potentials to improve WEEE management because of the educational sphere. As the generated WEEE amount will most likely increase further in the following decades and the current situation offers plenty of possibilities for improvement, future generations will have to deal with this topic. Therefore, possibilities to further educate people in pre-school, school, university and CPD-level should be considered at this station. Based on the stakeholders′ input to these four

topics, relevant information about the knowledge, handling and social interaction related to WEEE management in Brazil were to be derived.
Two moderators were steadily present at each station to introduce the defined topics to each group on the one hand and on the other to document every idea, content and output of the discussion. Considering possible language barriers in-between the group members, it was ensured that the moderators covered Portuguese, Spanish and English language skills. The moderators remained at their dedicated station. During the World Café a timekeeper was determined to ensure that every group changed topics after 15 minutes; the overall time of the Wold Café was 60 minutes. In total 44 participants out of 20 different institutions mainly from Brazil and Nicaragua participated in the World Café and presented their input to the outlined topics. In the framework of this paper the results of station 2 of the LaWEEEda World Café are analysed and discussed.

3 Results of the World Café Regarding Gaps and Needs in the Brazilian WEEE Management System

In the following sections the results regarding the main gaps and needs in the Brazilian WEEE management are visualized and discussed related to the stakeholder groups government, companies, scientists and consumers. As a direct outcome specific recommendations are derived. When evaluating the entire data collected, it is striking that many ideas and thoughts have already been formulated respectively documented as recommendations for the related group of stakeholders.

3.1 Main Gaps and Needs of the Brazilian Government Regarding WEEE Management

The following Table 3 visualizes the aspects mentioned by the participants of the World Café about the gaps and needs regarding the government related to the management of WEEE in Brazil.

Table 3. Gaps and needs of the government regarding Brazilian WEEE management in the order they were stated

Government	
Execution of national policy of solid waste	Incentives for correct disposal
Incentives to eco design	Encourage entrepreneurship. Professionalize the business
Communicate collection points	To make partnership and cooperation agreements with all stakeholders
Improve fiscalization on WEEE law	Companies need to pay for the collection of waste
Assistance with the cooperatives (infrastructure and financial resources)	Sanctions and incentives for whoever does or does not discard correctly
Structure and consolidate the market for WEEE	Technical, legal and management advice
Develop database on WEEE	Educate consumers
Agility to implement sectoral agreements	More collection points
Motivation on WEEE management	Inclusion of all stakeholders for the effective implementation of the national policy

According to the participants, the main gaps and needs are linked to lacking incentives for proper WEEE treatment and weak enforcement of the national policy for solid waste. In addition to that it would be helpful to reduce the bureaucratical effort to facilitate actions for all stakeholder groups. A database is mentioned as being useful to inform interested parties about the current situation in WEEE management. Due to higher transparency a further network of cooperatives can develop and accelerate development in WEEE treatment processes in different locations of Brazil. Linked to this, an open access platform to inform consumers about appropriate WEEE treatment is considered to be relevant for improving the overall WEEE management. By explaining shared responsibilities, in which the consumer is among others responsible to return EEE to retailers, a first step to improvement might be achieved. An open online platform with current data and basic information would additionally be useful to create a closer network in the frame of WEEE management in Brazil.

3.2 Main Gaps and Needs of Brazilian Companies Regarding WEEE Management

The following Table 4 shows the mentioned ideas and thoughts of the participants in the World Café regarding the main gaps and needs of companies for Brazilian WEEE management.

Table 4. Gaps and needs of companies regarding Brazilian WEEE management in the order they were stated

Companies	
Incentives eco design	More collection points
Companies with legal requirements	Incentives to correct disposal
Assistance with the cooperatives (infrastructure and financial resources)	Encourage entrepreneurship. Professionalize the business
Structure and consolidate the market for WEEE	To make partnership and cooperation agreements with all stakeholders
Integrate and consolidate cooperatives (network)	Companies need to pay for the collection of waste
Professional qualification for all processes	Sanctions and incentives for who dies or does not discard correct
Develop efficient technologies	Technical legal and management advice
Develop data base on WEEE	Improve current management systems
Downsizing of the recycling process	Improve understanding of national solid waste policy
Motivation on WEEE management	Educate the consumer
Inclusion of all stakeholders for the effective implementation of the national policy	

First of all, it is perceptible that many thoughts and ideas are identical to those mentioned for the governmental gaps and needs (see section 3.1). Creating a proper network in-between stakeholder groups in the WEEE business would be relevant to improve the overall WEEE management practice. Technical, legal and management advice should be provided to ensure sustainable WEEE treatment in Brazil from the companies' perspective. From companies the participants wish for more technical innovations and a motivation for proper WEEE management. This considers also product design, which should be eco-friendlier than in today's practice. As already mentioned in the governmental section, the education of the consumer is considered as a key instrument for improving WEEE treatment. If every stakeholder knew how to handle WEEE properly and why it is relevant to do so, a major step to enhanced WEEE treatment schemes could be marked.

3.3 Main Gaps and Needs of Brazilian Scientists Regarding WEEE Management

The following Table 5 builds upon the ideas and thoughts of the participants of the World Café about the gaps and needs of the scientists in the field of WEEE management respectively treatment.

Table 5. Gaps and needs of scientists regarding Brazilian WEEE management in the order they were stated

Scientists	
Disciplines focused on WEEE	Encourage entrepreneurship. Professionalize the business
Development programs on WEEE to society	Development of more research and extension activities on WEEE and recycling business
Report on waste categorization	To make partnership and cooperation agreements with all stakeholders
Professional qualification for all processes	Network of cooperatives to achieve economies of scale
Develop of efficient technologies	Technical, legal and management advice
Environmental management as a compulsory subject	Educate the consumer
Motivation on WEEE management	Demystify negative concept of corporatism
More collection points	Inclusion of all stakeholders for the effective implementation of the national policy

For scientists in the business of WEEE the participants of the World Café identify possibilities related to the development of more efficient technologies and programs for society. They are further asked to educate the consumer by giving technical, legal and management advice regarding sustainable treatment of EEE. Gaps and needs are further identified in the lack of a report on waste categorization. This aspect could be implemented in the before mentioned online platform (see section 3.1) to provide "first-hand" information to the consumer and every other stakeholder group. In addition to that, the scientists are called to deepen their research and to extend activities on WEEE respectively its recycling.

3.4 Main Gaps and Needs of Brazilian Consumers Regarding WEEE Management

The following Table 6 shows the suggestions mentioned by the participants within the World Café regarding the gaps and needs of consumers in the Brazilian WEEE management system.

Table 6. Gaps and needs of consumers regarding Brazilian WEEE management in the order they were stated

Consumer	
Consumer awareness	More collection points
Communicate collection points	Incentives for correct disposal
Report on waste categorization	Encourage entrepreneurship, professionalize the business
Motivation on WEEE management	Demystify negative concept of corporatism
Inclusion of all stakeholders for the effective implementation of the national policy	

The main gaps identified by the participants of the World Café related to the consumers are linked to those of other stakeholder groups. As in the previous sections related to other stakeholder groups, incentives are mentioned which should motivate consumers to dispose of their WEEE correctly. Further gaps are outlined in missing education. With a higher number of implemented collection points, ideally communicated well to the consumer, the awareness of the importance to dispose of WEEE correctly might be affected positively. The main recommendation in this field is to include all stakeholder groups for the effective enforcement of the already implemented national solid waste policy. If synergies are activated and the stakeholder groups cooperate closely, a reduction of gaps and needs can be reached which would directly affect the overall WEEE management in a positive manner.

4 Conclusion

Increasing WEEE generation and its proper treatment is a global issue. As recycling systems are inefficient in LATAM, including Brazil, it is important to analyse the main gaps and needs to start to work on. Through the methodology of a World Café the main gaps and needs in the Brazilian WEEE management were pointed out and allocated to the stakeholder groups government, companies, scientists and consumers. Many ideas considered by the participants are linked to closer interactions between the stakeholder groups. Applying synergies and cooperating closely on the path to increasing the efficiency of WEEE management schemes in Brazil is therefore the main recommendation. As no reliable database exists, stakeholders, especially the consumer, need better education and reliable sources providing information about how to handle WEEE properly after using it, and an open access platform would be reasonable for implementation. Such a platform could also lead to better control instruments for the government, which would directly expand its opportunities to enforce the officially implemented national policy of solid waste in Brazil. Scientists are asked to provide reliable data and invent more efficient technologies regarding WEEE treatment. Within all stakeholder groups more incentives are required to raise the motivation for correct disposal of WEEE.

Brazil is confronted with many challenges regarding the proper treatment of still rising amounts of WEEE. But with close interaction throughout all stakeholder groups, at least great contributions can be made. On the pathway to reaching the SDGs including all its targets in the frame of the agenda 2030, various sectors have to be improved on the global scale – the WEEE sector is definitely one of them.

5 Zusammenfassung

Elektro- und Elektronikaltgeräte (engl. "waste electrical and electronic equipment" (WEEE)) und deren Verarbeitung sorgen weltweit für große Herausforderungen. Die Behandlung von WEEE erfordert daher auf politischem und umweltregulatorischem Druck basierende präventive Maßnahmen. Innerhalb Lateinamerikas (LATAM) produziert Brasilien mit steigender Tendenz die größten Mengen an WEEE, Recyclingprogramme stecken dagegen noch in den Kinderschuhen [1]. Es ist daher wichtig, die Lücken und Anforderungen vergleichbarer Systeme aufzuzeigen. Im Rahmen dieses Beitrags werden Empfehlungen zur Verbesserung des WEEE Managements in Brasilien gegeben, die auf dem Wissen verschiedener lokaler Experten basieren. Die Ergebnisse der Wissensaustauschmethode „World Café" zeigen, dass die politische Durchsetzung nationaler Gesetze und eine enge Kooperation verschiedener Stakeholdergruppen Schlüsselinstrumente zur Verbesserung des WEEE Managements in Brasilien darstellen.

Acknowledgments. This research was supported by the Erasmus+ funded project "Latin American-European network on waste electrical and electronic equipment research, development and analyses" (LaWEEEda). We thank our colleagues from Asociacion Nicaragua Ambiental (Ni), Asociacion Universidad Cristiana Autonoma De Nicaragua (Ni), Asociacion Universidad Tecnologicala Salle (Ni), Associacao Dos Catadores Do Aterrometropolitano Do Jardim Gramacho (Br), Cooperativa Popular Amigos Do Meio Ambiente Ltda (Br), Hanon Tercero Metales Y Compania Limitada (Hanter Metals. Ltd) (Ni), The University Of Northampton (Uk), Universidade Estadual Paulista - Unesp (Br) and Universidade Federal Do Rio De Janeiro (Br) who provided insights and expertise that greatly assisted the research, although they may not agree with all of the conclusions of this paper.

6 References

[1] Federico Magalini, Ruediger Kuehr and Cornelis Peter Baldé 2015 eWaste in Latin America: Statistical analysis and policy recommendations
[2] Vereinte Nationen 2015 Transformation unserer Welt: die Agenda 2030 für nachhaltige Entwicklung
[3] Cucchiella F, D'Adamo I, Lenny Koh S C and Rosa P 2015 Recycling of WEEEs: An economic assessment of present and future e-waste streams vol 51
[4] Baldé K, Wang F, Kuehr R and Huisman J 2015 The global e-waste monitor 2014: Quantities, flows and resources (Bonn: United Nations Univ., Inst. for the advanced study on sustainability)
[5] Salhofer S and Tesar M 2011 Assessment of removal of components containing hazardous substances from small WEEE in Austria vol 186
[6] Dias P, Machado A, Huda N and Bernardes A M 2018 Waste electric and electronic equipment (WEEE) management: A study on the Brazilian recycling routes vol 174
[7] Baldé C P, Forti V., Gray V, Kuehr R and Stegmann P 2017 Global-E-waste Monitor 2017: Quantities, Flows, and Resources vol 2017 (Bonn/Geneva/Vienna)
[8] Acosta E 2018 Latin America could be 'the new frontier' in scrap processing https://recyclinginternational.com/non-ferrous-metals/acosta-latin-america-could-be-the-new-frontier-in-scrap-processing/
[9] Marshall R E and Farahbakhsh K 2013 Systems approaches to integrated solid waste management in developing countries vol 33
[10] Souza R G de, Clímaco J C N, Sant'Anna A P, Rocha T B, do Valle R d A B and Quelhas O L G 2016 Waste Management **57** 46–56
[11] Brown J and Isaacs D 2005 The World Café: Shaping our futures through conversations that matter 1st edn (San Francisco, Calif.: BK Berrett-Koehler Publishers)
[12] Chang W-L and Chen S-T 2015 The impact of World Café on entrepreneurial strategic planning capability vol 68
[13] Carter E, Swedeen B, Walter M C M and Moss C K 2012 Research and Practice for Persons with Severe Disabilities **37** 9–23
[14] Fouché C and Light G 2011 An Invitation to Dialogue: 'The World Café'In Social Work Research vol 10
[15] Hodgkinson G P, Whittington R, Johnson G and Schwarz M 2006 The Role of Strategy Workshops in Strategy Development Processes: Formality, Communication, Co-ordination and Inclusion vol 39
[16] Johnson G, Prashantham S, Floyd S W and Bourque N 2010 The ritualization of strategy workshops vol 31
[17] Schieffer A, Isaacs D and Gyllenpalm B 2004 The world café: part one vol 18

Formal and Informal E-waste Collection in Mexico City

Nina Tsydenova[1] and Merle Heyken[1]
[1]Department of Ecological Economics, Carl von Ossietzky University of Oldenburg, Oldenburg, 26129, Germany
nina.tsydenova@gmx.de; merle.heyken@gmx.de

Abstract

Cities in emerging countries, which attract the rural population with their higher living standards, cause various problems affecting the environment. One of them is the fast increasing amount of waste electrical and electronic equipment (WEEE), known as "e-waste". One of the main prerequisites for a sustainable e-waste management is the separate collection. This study discusses the collection system of e-waste in Mexico City, both formal and informal. The formal collection includes the separate collection of e-waste according to the new separation system. In addition, the government is running two other initiatives, the Barter Market and Reciclatrón, which represent collection sites for e-waste. However, these efforts are not sufficient, which is why the informal sector plays a significant role in collection, recycling and disposal of e-waste.

1 Introduction

In emerging countries, rapid urbanization as cities attracts the rural population with their higher living standards causes various problems affecting the environment. One of them is the fast increasing amount of waste electrical and electronic equipment (WEEE), known as "e-waste".

Most electronic devices have a useful life pre-established by the producer. After tahe specific time, the equipment should be replaced as defined by the product life cycle. Setting an end to the functional period makes the consumers buy new products. The objective of this practice is to assure a circulation of merchandise, gaining more profits, while the environmental issues are not considered [1]. Therefore, the society in developing economies as well as in industrialised countries accelerates the frequency of replacement of electrical and electronic equipment. Moreover, technological progress also influences the promotion of obsolescence of electronic equipment.

Obsolescence is a concept that considers the influence of marketing and consumption culture to replace one's electronic equipment. It appeals to the desire of customers to have a better lifestyle. This practice is associated with three types of obsolescence promoted by manufacturers: functional obsolescence, when a product substitutes another with a better performance; technical, when a new product substitutes the older version; and style, which occurs when a completely functional electronic device stops being desired because of popular fashion. This pressure on the users directly influences the high generation of e-waste [2].

It is important to highlight that technological obsolescence periods are becoming shorter, and the acquisition costs reflect a downward trend. Technological obsolescence is an incremental phenomenon for our information- and knowledge-based society. The Electrical and Electronic Engineering (EEE) industry is responsible for 10%–20% of the global environmental impact related to the use of non-renewable resources [3]. In 2016, around 44.7 million tons of WEEE were generated worldwide [4]. In response to this challenge, decision-makers worldwide are searching for a sustainable way of handling the e-waste. A recent study describes the solutions to the e-waste problem in Mexico City. The city generates around 12 000 Megagramm (MG, 1 MG corresponds to 1 metric ton) per day, of which 312 MG is represented by e-waste [7].

In this study, the e-waste includes all discarded electrically powered appliances, such as computers, televisions, electronic games, photocopiers, radios, video recorders, DVD players and cell phones, and traditionally non-electronic goods, such as refrigerators, washing machines, dishwashers, and ovens. E-waste includes both 'white' goods (e.g., refrigerators, washing machines, and microwaves) and 'brown' goods (e.g. televisions, radios, and computers) that have reached the end of their life [5]. This corresponds to the WEEE definition of the European Union, which includes household appliances, IT and telecommunications equipment, lighting (except household bulbs), electrical and electronic tools, toys, leisure and sports equipment, medical devices (except implanted and infected products), monitoring and control instruments, automatic dispensers. In general, the WEEE does not include large-scale industrial tools [6].

Latin American countries have adopted the technology of industrialised countries. Latin America generates 9% of the whole WEEE worldwide. Of this, Brazil and Mexico produce the highest amount of WEEE in the region. In 2016, Brazil generated 1534 kg of e-waste and Mexico 998 kg, respectively [4]. Figure 1 gives an overview of the percentage of e-waste in the municipal solid waste (MSW) composition in Latin American countries. Mexico has an elevated level of e-waste generation for the region, as e-waste represents 2.6% of MSW [7].

This huge consumption of electronic devices helps companies to gain more customers in a rapidly evolving market. These consumer trends trigger serious economic, social and environmental problems because of a lack of public policies aimed at sound management of e-waste. When the electronic equipment has completed its life cycle, it becomes e-

A. Pehlken et al. (Eds.), *Cascade Use in Technologies 2018*,
https://doi.org/10.1007/978-3-662-57886-5_5

waste, and is in most cases kept opencast without any treatment, even though monitors, keyboards, cable, circuits and drives of computers generate a large volumes of toxic waste. For instance, because of a lack of proper waste management in Mexico e-waste can end up in open dumps, which still exist in the country.

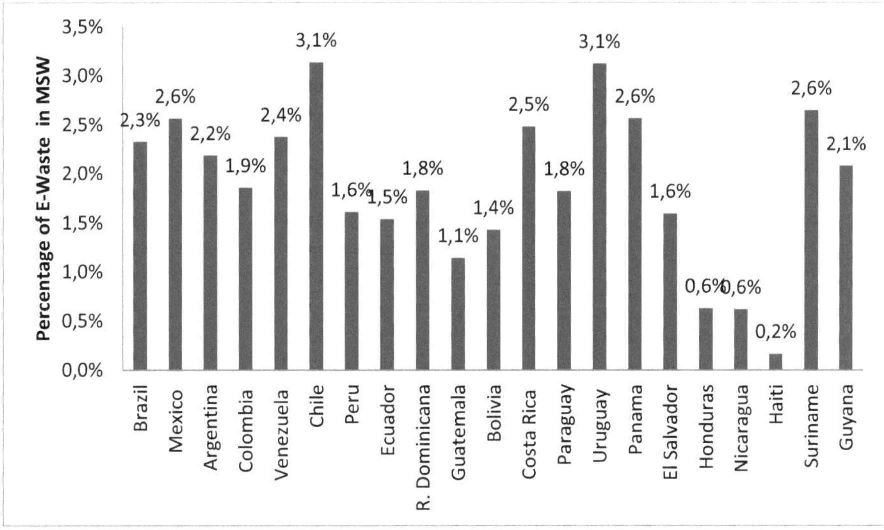

Figure 1. Percentage of e-waste in the MSW composition in Latin America (based on [7] and [8])

It is estimated that 50% of the e-waste weight corresponds to metals. Steel, aluminum, copper and precious metals are the main types. The rest of the materials (plastic and glass) are distributed in similar percentages. Electronic waste also contains high value metals like gold, silver and copper that can be recovered and brought back into the production cycle. In some studies it is mentioned that with appropriate treatment between 70–90 % of e-waste could be recycled or reused [9]. This allows electronic waste to be a source of income, however, in most countries of Latin America, including Mexico, e-waste just occupies space in the landfills instead of being recycled.

The lack of mechanisms for collection and appropriate treatment, and improper handling and management of toxic substances associated with e-waste harm human health and disrupt ecosystems. E-waste can contain lead, mercury and cadmium, all of which are extremely toxic. Up to 6% of the weight of old computer or television monitors may be lead, the main part of which is located in the Cathode Ray Tube [10]. According to Ramírez and Escalera, the main environmental pollutant of all e-waste types are the computer devices (including the batteries and peripheral products) because of their toxic content and their mismanagement, causing damages to the environment [11]. Therefore, it is essential to apply chemical and physical treatments to prevent toxic emissions to the environment. An important prerequisite for the implementation of adequate processing schemes for e-waste is the efficient separation at the source and applicable separate collection. Therefore, it is important to implement separate collection and treatment systems for e-waste. This study gives an overview of collection methods for e-waste applied in Mexico City, both informal and formal. The formal methods include the municipal separate collection of e-waste and collection sites organised by the government. All other activities, which complement the official collection system, are considered informal.

2 State of the Art

Mexico is a diverse country with 125 million residents and abundant natural resources. Mexico is a member of the Organization for Economic Co-operation and Development (OECD) and simultaneously a developing country with a GDP per capita of US $ 8201.3 [12]. This study focuses on Mexico City, the capital and the most populated city. Mexico City is an increasingly globalized and spatially growing city in which waste management is a critical issue in terms of urban and environmental governance.

Aleman [10] sees the main problem associated with e-waste in the absence of a proper regulation of collection and treatment. In Latin America, there are only a few countries that have a defined regulatory framework and can count on formal recycling systems. However, they are often at an initial phase and improvements need to be done in the whole region. As reported by the UN, Mexico collects most of the e-waste in Latin America, approximately 36% of the e-waste generated [4]. However, only 17 % of the produced WEEE is recycled [14].

The legal framework for waste management is based on the Mexican Political Constitution, particularly article 115, which states that municipalities are responsible for providing public cleaning services, as well as being in charge of collecting, transporting, treating, and final disposal of waste. However, only 19 out of 32 states have legislation on waste management. These states are: Aguascalientes, Baja California, Chiapas, Chihuahua, Federal District, Durango,

Guanajuato, Guerrero, Hidalgo, Jalisco, Michoacán, Nuevo León, Puebla, Querétaro, Quintana Roo, Sonora, Tabasco, Tamaulipas and Veracruz [15].

2.1 E-waste in Mexico City

Mexico City, as other cities in the developing world, faces the challenge of having effective strategies of E-waste management. Some of the main problems are the following: lack of infrastructure for pre-process and treatment, absence of facilities and technologies of innovation, lack of investment, high management costs [2].
Several important steps were made in Mexico City in response to this challenge. In 2018, the City Council of Mexico City introduced the cascade use concept (reduce, reuse, recycle) into the general waste law of the city, which was also renamed into the Law of Integrated Waste Management of Mexico-City [16].
According to the new waste law of Mexico City, NADF-024-AMBT 2013, which came into force in 2017, e-waste is considered as a category of special waste, which should be collected separately. Special waste is to be collected on Sundays. However, due to a poor management, e-waste management does not function properly. Because of an absence of waste containers in the city, citizens have to bring their refuse directly to the waste trucks. However, trucks do not follow any schedule, which makes it difficult for the population to follow the regulations. This way, e-waste is delivered to the trucks on other days, consequently it is mixed with other fractions. In addition, household waste may be given to the waste workers who sweep the streets and bring waste to the truck themselves or directly to a transfer station.
Health is one of the problems associated with recovery activities in Mexico. The main reason is that a large share of the recycling activities is performed by the informal sector. Workers involved in the informal sector do not use appropriate equipment and have little or no knowledge about potentially hazardous elements. This lack of knowledge and equipment leads to the exposure to health problems of not only informal WEEE-related workers but also the public [2].
Apart from this, Mexico City has the problem of scarce landfill capacities since the local landfill was closed in 2011. After that, the city has had to send its waste to the nearby disposal areas, which are located in the closest federal states. Consequently, the transportation costs increased significantly. It is assumed that the transportation costs make up the biggest part of waste management costs in Mexico City (16.70 Euros per MG of inorganic waste) (1 € equals 23.97 Mexican pesos as of 18.06.2018) [17]. Due to the poor recovery, the e-waste is sent to the landfills together with unrecyclable litter, which cause an increase of the waste management costs for the city and a shortage in landfill capacity.
In Mexico, full WEEE recycling processes are still not common. Currently, the amount of recovered materials is very small, and the political framework and infrastructure are still limited. Therefore, the existing infrastructure cannot process large amounts of WEEE. Most of the WEEE recycling companies do not offer a full recovery cycle because they focus only on the recovery of valuable components, leaving non-valuable components aside. Those non-valuable components still represent a threat to the environment and public health [2].
The separate collection is important for the solution of the e-waste problem. Today, 50% of generated e-waste ends up in landfills, while 40% remain in the households or warehouses [18]. To change this situation, the city government has started several campaigns ,the results of which are going to be discussed in this study. In order to raise awareness and promote waste separation, the government organizes Barter Markets. At this event the citizens can exchange the separate waste fraction for food products, such as vegetables, dairy products and seeds. Apart from other waste fractions, it is also possible to exchange WEEE for nutrition products. Another campaign, called Reciclatrón, targets e-waste directly. The participants of Reciclatrón have the chance to bring WEEE to the special collection site. As a reward for their environmentally friendly behavior, the participants receive pot plants and a bag of compost that they can utilize for their gardening. The collected waste is sent to the company Cali Resources located in Tijuana, Baja California, for recycling [19].
Apart from recycling, reuse is quite common in Mexico. This activity is much more private than recycling; therefore, people do not depend on the existing infrastructure created by the government. In rural communities, for example, discarded cans, cardboards and glass are often reused in the household. In the city where products are more accessible, this practice is not as popular. Nevertheless, people reuse items in different ways. They use empty bottles as flower pots or storage containers. Reuse of clothes is also broadly accepted. Since the number of family members is high, one piece of clothing can be used by 3-4 children. Other products that are often reused in Mexico are steel components of cars, bicycles and furniture [20]. The study of Corral-Verdugo [20] showed that reuse and recycling are complementary activities in the households. For example, aluminum cans are normally not reused because they are recycled in most of the cases, while plastic bags are never recycled but often reused. However, items such as newspapers and paper are hardly recycled or reused (despite the presence of a paper recycling facility in the region).

3 Case Study Description and Assessment

The waste management system of Mexico City shows different pathways for each waste fraction. This study schematizes the collection of e-waste in Mexico City. The estimation of the rate of collection of e-waste is based on the official data presented by the SEDEMA (environmental authority of Mexico City) and other literature sources.

Separate waste collection represents the prerequisite for reuse, recycling and recovery of raw materials and provides the basis for the use of waste as an economic source. Therefore, the proper separation and collection play an important role in the recycling of e-waste [21]. This section discusses collection practices of Mexico City targeting e-waste.

3.1 Formal Collection

Barter Market

The first official collection program to be discussed is the Barter market (Figures 2 and 3). This initiative started in 2012 based on the environmental protection law of the city. The program takes place every month in one of the sightseeing places of the city. The main purpose of the event is to educate the citizens to separate the waste in their households.

Figure 2 and **Figure 3**. Barter market in May 2017 [22]

In order to participate in the program, the citizens have to bring 1-10 kg of clean separated waste fractions. For the specific waste fractions, the participants get a different amount of points. Table 1 presents the number of points which the participants can earn. The points can be exchanged for food products. The number of points given for e-waste depends on the category: electronics (group A), cables (group B) or mobile devices (group C). However, it should be mentioned that the number of points given for e-waste is significantly lower than for packaging waste, such as Tetra Pak and PET bottles [23]. This focus on PET bottles can be explained by the fact that Mexico has the highest consumption of bottled drinks in the world [24], which leads to a higher generation of wasted PET bottles.

Table 1. Number of points earned for each category [25]

Waste category	Aluminum	Cardboard	Cans and iron	Plastic HDPE	Glass	Paper	Tetra Pak	PET	E-waste		
									A	B	C
Number of points	2					3	8	18	4	2	5

In 2016, 109.68 MG of separated waste fractions were collected in exchange for 47 food products. 24 local producers and 36606 assistants helped with the organization of the event. It should be mentioned that only 7% of the collected recyclables are represented by e-waste, while in 2015 WEEE was the second biggest waste fraction collected. In comparison to 2015, the amount of collected e-waste in 2016 dropped significantly by 16 MG. This fluctuation can be explained by the fact that the generation of e-waste is not as fast as that of other categories, such as the packaging waste. Overall, during all the editions of the Barter market 77.4 MG of e-waste were collected, making up 11% of the collected recyclables since 2012 (Figure 4). The modest results obtained through the Barter market can be explained by the lower exchange rate of e-waste and competing collection programs such as Reciclatrón and informal collection.

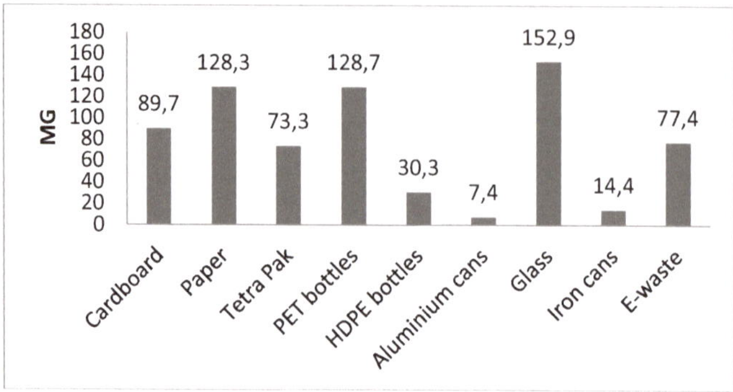

Figure 4. Cumulative quantity of collected waste fractions at the Barter market since 2012 [25]

Reciclatrón

The Reciclatrón (Figures 5 and 6) was organized for the first time in 2013. As the Barter market, Reciclatrón is held each month. It typically takes place in the location of one of the main universities in Mexico City. During Reciclatrón everyone can bring their e-waste in exchange for the compost and pot plants. In 2016, 273.3 MG of WEEE were collected even though the rewards are not particularly attractive, as private gardening is not common in Mexico City.

Figure 5. Reciclatrón at University UAM Azcopotzalco. February 2017

Figure 6. Box for collection of cell phones at Reciclatrón at University UAM Azcopotzalco. February 2017

This program defines five categories of e-waste, while in the Barter market there are only three categories. Table 2 presents the categories of WEEE applied at Reciclatrón. It should be mentioned that the category E that includes monitors, TVs, electrical ballasts, screens, transformators, lamps, heating devices, batteries, toners and refrigerators, was introduced first in 2016. Nevertheless, this category was the largest to be collected and represented 76% of the total sum [25].

Table 2. Categories of e-waste at Reciclatrón [25]

Category	Description
A	Printers, copy machines, keyboards, mice, calculators, cameras, typewriters, faxes, radios, voice recorders, devices of uninterruptible power supply, microwaves, vacuum cleaners, blenders, dishwashers, coffee machines, DVDs, VHS, MP3, video games, amplifier, PDAs, microcomponents, fixed lines, projectors, equalizer, flat irons, speakers, hair dryers.
B	CPUs, laptops, hard disks, cards for computers.
C	Mobile phones.
D	Charger, cables, motors.
E	Electrical ballasts, monitors, alkaline batteries, transformators, lamps, heating devices, betteries, refrigirators, toner cartridges

Reciclatrón showed a rising trend during the last years. The amount of WEEE collected grew by 178.02 MG from the first edition in 2013 (Figure 7) [25].

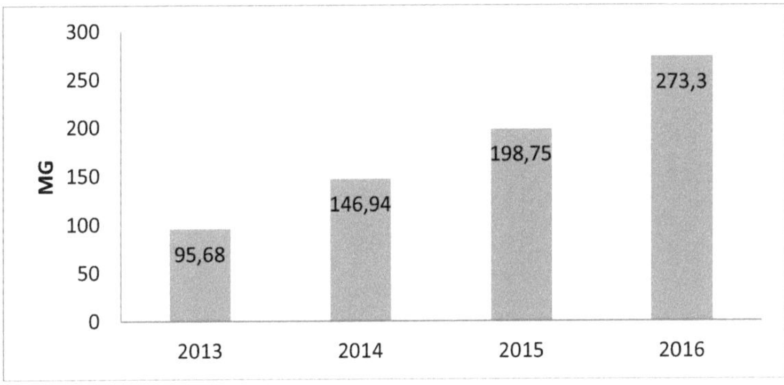

Figure 7. E-waste collected during Reciclatrón 2013-2016 [25]

Municipal Collection of Household Waste

New waste law norm NADF-024-AMBT-2013 was introduced in July 2017. This aims to increase the amount of collected recyclable materials and thereby decrease the quantity of landfilled waste. Since 2011, municipal solid waste in Mexico City has been separated in households into organics and inorganics, while the new law Norma NADF-024-AMBT-2013 mandates the new segregation of residual waste into 4 fractions: organic, recyclables, non-recyclables, hazardous and bulky waste [26]. E-waste is included in the category of hazardous waste and should be collected on Sundays. However, collection of municipal solid waste (MSW) represents one of the biggest challenges in waste management in Mexico City. The staff of a waste truck normally consists of a driver and several volunteers, who help to sort the waste and are part of informal recycling. They normally break the trash bags which the citizens bring to them and sort the content themselves searching for the recyclables that they sell later. This practice of the waste workers does not stimulate the population to perform proper separation. Although the results of the influence of the new law on the efficiency of separate collection are not available so far, it can be assumed that the separation rate is not going to increase. In 2016 the separation was only 34% under the condition of 2-bin-separation [25].

3.2 Informal Collection

Over the past 60 years, Mexican cities have grown rapidly, mostly due to migration from rural areas. Many migrants survive by engaging in informal waste sector activities. The informal sector plays a significant role in the Mexican economy, employing about 20 million people [27]. Urbanization often leads to the creation of new slums. Individuals who live in squatter settlements often do not pay taxes. Municipal solid waste consumes 20–40% of municipal budgets [28]. This low level of taxation translates into a low availability of funds for municipalities. This gap in provision of waste services is normally filled by the informal sector. Mexico City generates over 12,000 MG of MSW a day. For comparison, the whole country of Greece generates 15,054 MG of MSW per day [8]. 67% of MSW from Mexico City are landfilled [25]. A complex informal system has developed around MSW collection, recycling and disposal over the past 60 years. Approximately 20,000 individuals make a living in the informal sector around residues [29]. The informal sector collects around 7% of the e-waste generated in Mexico City [30]. No health and safety standards are followed by the informal sector. Even though the income and living conditions of informal waste workers differ significantly depending on their main activities, the majority of informal waste workers (dump and street waste pickers) constitute the lowest level of society. Working conditions include permanent exposure to dangerous and toxic substances. Waste pickers are subjected to harassment from officials, exploited by traders of recyclables and have no legal protection. They live in humiliating circumstances and generally lack sanitary services, health care and social benefits.

Recoverable e-waste materials are sold according to the value in the local domestic market. Based on the data in Table 3, the greatest economic resources are obtained from motherboards, scrap copper cables [18].

Table 3. Market price of recovered materials [18] (1 € equals 23.97 Mexican pesos. as of 18.06.2018)

Material	Commercial value (€ per kg)
Plastic	0.02
Scrap	0.09
Keyboards	0.04
Cables	1.07
Cooper	3.01
Aluminum	0.8
Motherboard	2.15
Hard Disk	0.21

4 Conclusion

Mexico City is on its way to developing efficient e-waste collection systems. The city is advancing the source separation through the introduction of new legal norms. If the efficiency rate increased, this would contribute a lot to the recovery of bigger numbers of fractions. However, the source separation in Mexico City does not function well and the separation efficiency is very low. For a sustainable e-waste management, it is important that users know how and where to deliver obsolete equipment. For the functioning of the system, it is vital to establish collection sites with basic standards of environmental protection and health and safety [2]. Therefore, the city has developed other initiatives such as the Barter Market and Reciclatrón to promote a better separation and collection of WEEE. Overall, the waste collection campaigns are successful in raising public awareness, but the collection rates for e-waste of both initiatives are low. The Barter Market allows the collection of 0.007% of generated e-waste, while Reciclatrón achieves 0.24%. The food products provided in the case of the Barter Market are more attractive for citizens than the compost and pot plants at the Reciclatrón. Therefore, it is suggested to offer food products as a reward for the environmentally friendly behavior in both campaigns in order to increase the collection rate. In addition, in case of the Barter Market it may make sense to shift the focus from plastic waste to paper since the last fraction is easier to recycle due to its more homogeneous composition. However, there is almost no market for waste paper due to its low price.
Nevertheless, it is expensive to implement such campaigns, as they require high upfront investment. Therefore, the informal sector complements the official programs and participates actively in e-waste management. However, the urgent integration of the informal sector in the formal activities is needed to improve the working conditions and provide legal protection for the workers.
The management of e-waste is a global priority, therefore, Mexico City, as the capital and leader in environmental policy must develop and implement management plans aimed at electronic waste collection, characterization, quantification, recovery and reuse, as well as the commercialization of the components. The linked and coordinated coordination of all levels of government is needed, together with the informal sector to promote a sustainable approach to e-waste management.

5 Zusammenfassung

Großstädte in Entwicklungsländern ziehen aufgrund der besseren Lebensbedingungen die ländliche Bevölkerung an, verursachen jedoch verschiedene Umweltprobleme. Eines davon ist die ständig ansteigende Menge an Elektroschrott (engl. „waste electrical and electronic equipment" (WEEE)). Als Bedingung für ein nachhaltiges Management des Problems kann die separate Sammlung des Elektroschrotts angesehen werden. Dieser Beitrag diskutiert die formellen und informellen WEEE Sammelsysteme in Mexiko Stadt. Das formelle Sammelsystem beinhaltet hierbei die getrennte Sammlung von Elektroschrott im Rahmen des neuen Mülltrennungssystems. Zusätzlich fungieren zwei weitere staatliche Initiativen, „Barter market" und „Reciclatrón" als Sammelstätten für Elektroschrott. Da diese Bemühungen bisher nicht ausreichen, spielt der informelle Sektor weiterhin eine bedeutende Rolle bei Sammlung, Recycling und Entsorgung von Elektroschrott.

6 References

[1] Carretero, A (2015) ¿Avances en la prevención y reducción de residuos de aparatos eléctricos y electrónicos? CESCO de Derecho de Consumo 13: 214–222. https://previa.uclm.es/centro/cesco/pdf/trabajos/34/27.pdf
[2] Cruz-Sotelo, SE, Ojeda-Benítez S, Sesma J, Velázquez-Victorica KK, Santillán-Soto N, García-Cueto OR, Alcántara V, Alcántara C (2017) E-Waste Supply Chain in Mexico: Challenges and Opportunities for Sustainable Management. Sustainability 9(4): 503. https://doi.org/10.3390/su9040503
[3] Georgiadis, P, Besiou, M (2009) Environmental strategies for electrical and electronic equipment supply chains: Which to choose? Sustainability 1: 722–733. https://doi.org/10.3390/su1030722
[4] Baldé, CP, Forti V, Gray, V, Kuehr, R, Stegmann, P (2017) The Global E-waste Monitor – 2017, United Nations University (UNU), International Telecommunication Union (ITU) & International Solid Waste Association (ISWA), Bonn/Geneva/Vienna.

[5] Khetriwal DS, Kraeuchi P and Widmer R (2009) Producer responsibility for e-waste management: Key issues for consideration-Learning from the Swiss experience. Environmental Management 90: 153–165.
[6] Directive 2012/19/EU of the European Parliament and of the Council of 4 July 2012 on waste electrical and electronic equipment (WEEE) https://eur-lex.europa.eu/legal-content/EN/TXT/?uri=CELEX:32012L0019
[7] GSMA (2015) E-waste en America Latina. Análisis estadístico y recomendaciones de política pública. https://www.gsma.com/latinamerica/wp-content/uploads/2015/11/gsma-unu-ewaste2015-spa.pdf
[8] Hoornweg, D, Bhada-Tata P (2012) What a Waste: A Global Review of Solid Waste Management. Urban development series; knowledge papers no. 15. World Bank, Washington, DC. © World Bank. https://openknowledge.worldbank.org/handle/10986/17388 License: CC BY 3.0 IGO.
[9] Sánchez M, Bonales J, Espinoza R (2008) Contaminación del medio ambiente en la región oriente del estado de Michoacán por desechos electrónicos de equipo de cómputo obsoleto. Mundo Siglo XXI 13: 61-71.
[10] Alemán, CP (2017) Los residuos electrónicos un problema mundial en el siglo XXI. Cucyt Medio Ambiente 59 (13): 379-392.
[11] Ramírez G, Escalera ME (2018) Basura electrónica. Un estudio empírico en las PYMES. Congreso Virtual Internacional sobre Economía Social y Desarrollo Local Sostenible. https://www.eumed.net/actas/18/economia-social/22-basura-electronica.pdf
[12] World Bank (2016). Data Bank. http://data.worldbank.org/country/mexico
[13] AnnaMap (2018) Mexico Map. http://annamap.com/mexico/
[14] Alcántara-Concepción V, Gavilán-García A, Gavilán-García IC (2016) Environmental Impacts at the end of life of computers and their management alternatives in México, Journal of Cleaner Production. doi: 10.1016/j.jclepro.2016.04.125.
[15] Rojas L, Gavilán A, Alcántara V, Cano F (2012) Los Residuos Electrónicos en México y el Mundo. SEMARNAT-INECC.
[16] Valdez I (2018) ALDF modifica ley de residuos sólidos en la CDMX. http://www.milenio.com/df/aldf-modifica-ley-residuos-solidos-cdmx-biodigestion-termovalorizacion_0_1143486132.html
[17] Suarez G (2018) CDMX ahorra 597 millones de pesos por separación de residuos. http://www.eluniversal.com.mx/metropoli/cdmx/cdmx-ahorra-597-mdp-por-separacion-de-residuos
[18] Saldaña CE, Messina S, Rodriguez-Lascano Y, García M, Ulloa H (2016) E-waste in Mexico: a case study of Tepic, Nayarit. International the Conference on Waste Management and The Environment
[19] Valentini G (2018) Reciclatrón: jornadas de acopio de residuos electrónicos. http://greendates.com.mx/reciclatron-jornadas-de-acopio-de-residuos-electronicos/
[20] Corral-Verdugo V (1996) A structural model of reuse and recycling in Mexico. Environment & Behaviour 28, pp. 665 – 696. https://doi.org/10.1177/001391659602800505
[21] Agovino, M., Garofalo, A. & Mariani, A. Environ Dev Sustain (2017) 19: 589. https://doi.org/10.1007/s10668-015-9754-7
[22] La razón de Mexico (2017) Llega el Mercado de Trueque a Chapultepec, donde la basura cobra valor. https://www.razon.com.mx/llega-el-mercado-de-trueque-a-chapultepec-donde-la-basura-cobra-valor/.
[23] SEDEMA (2016) Inventario de residuos sólidos 2016. www.sedema.cdmx.gob.mx/storage/app/media/IRS-2016.pdf
[24] Plastics insight 82016) Global Polyethylene Terephthalate (PET) Resin Market. https://www.plasticsinsight.com/global-pet-resin-market/
[25] SEDEMA (2016) Inventario de residuos sólidos 2016. www.sedema.cdmx.gob.mx/storage/app/media/IRS-2016.pdf
[26] SEDEMA (2015) NADF-024-AMBT-2013. http://data.sedema.cdmx.gob.mx/nadf24/images/infografias/NADF-024-AMBT-2013.pdf
[27] Zúñiga J (2003) La informalidad es ya la principal fuente de empleo en la era Fox. La Jornada, Mexico City
[28] Medina M (2005): Serving the unserved: informal refuse collection in Mexico. Waste Management & Research 23(5):390 - 397
[29] Castillo H (1990) La Sociedad de la Basura: Caciquismo Urbano en la Ciudad de México. UNAM, Mexico City
[30] Alcántara-Concepción V, Gavilán-García A, Gavilán-García IC (2016) Environmental impacts at the end of life of computers and their management alternatives in México. J. Clean. Prod 131: 615.

E-Book Reader Recyclability

Ersin Karadeniz[1], Christian Klinke[1] and Jörg Woidasky[1]
School of Engineering, Pforzheim University, Pforzheim, 75175, Germany
joerg.woidasky@hs-pforzheim.de

Abstract

In Germany, there are currently about ten million e-book readers in use. Dismantling trials have been undertaken to assess the recycling potential of these devices. Ten discarded e-book readers from five different manufacturers were acquired for manual dismantling, out of which nine have been actually dismantled, whereas the tenth sample was reactivated by hard reset. During the trials the e-book readers were tested on various dismantling options such as required time, connection accessibility, destruction during dismantling and materials identification.

The major components of the e-book readers consist of the front and rear covers, the display system including the display unit, a mounting plate, the display cover, circuit boards and the accumulator. The arithmetic mean mass of all devices of this study was 207.1 g, including three products acquired without an accumulator. The arithmetic mean mass of the devices which were acquired in a complete state (samples 1-3 and 6-9) was 203.7 g before dismantling.

A dismantling difficulty rating scale was applied which focused on three different aspects (difficulty, accessibility, destruction), thus providing three to four rating stages for the dismantler. Moreover, X-ray fluorescence (XRF) analysis and Fourier-transform infrared spectroscopy (FTIR)- Attenuated total reflection(ATR) analysis enabled materials identification besides relying on the producers' materials marking of the polymer parts. As a result, the mean e-book reader composition was found to be 30 % polymers (PC, PC/ABS, PMMA, PET), 17 % metals (Zn, Al, Fe), 19 % glass, 14 % printed circuit boards, 15 % accumulator and 3 % of other materials such as cables and connectors. Besides the aforementioned fractions, the largest single material share was PC (13%) and aluminum (12%). Identification of the cumulative dismantling times of single, main components yielded arithmetic mean values of about 110 seconds for the accumulator, 155 seconds for the printed circuit board, and 445 seconds for the dismantling of the display, with a mean total dismantling time of 592 seconds. Due to the length of the dismantling time, apart from some zinc parts, manual dismantling is not economical. Even for metal parts, the KE_M calculation according to VDI 2243 showed only a slight indication of economical recycling. This is mainly caused by the high number of connections which needed to be opened (arithmetic mean values were 11.9 screws and 26.9 clips to be opened, using 5.6 different tools for these operations for one device). Typical dismantling characteristics were 10.7 dismantling steps per device, with an arithmetic mean duration of 53.4 seconds per dismantling step. Thus from a purely economical perspective, mechanical recycling should be preferred to manual dismantling for recycling.

1 Introduction

In 2016, 9.34 million individuals in Germany owned an e-book-reader, which is equal to about 11.3% of the German population. In 2013 there were only 5.59 million e-book-reader owners, indicating a 67% increase in a three-year period, and the market potential of these electronical devices has not been exhausted: More than two million customers are planning to acquire an e-book (Table 1) [1]. This indicates that these devices have occupied a significant segment in the electronics sector. Due to the low energy demand needed by the e-ink technology, e-book readers possess a specific design which differs from products such as notebooks or smartphones. Moreover, market development figures such as the decrease in the reader stock of 0.37 million devices from 2015 to 2016 have led to the expectation that significant amounts of obsolete e-book readers will become part of the waste electrical and electronical equipment (WEEE) flow in the near future. This is partly due to the substitution of e-book readers by devices such as larger smart phones, or tablet computers.

These statistics, along with the collection and recycling quota of the German law on WEEE (ElektroG [2]) which was increased recently, was the reason to work on e-book readers' recyclability assessment. In detail, re-use of the devices was tested, and the effort for dismantling both to meet legal requirements and to recover valuable materials was measured. Moreover qualitative information about design for recycling implementation was collected.

Table 1. E-book reader market in Germany [1]

Year	2013	2014	2015	2016
E-book reader owners (million)	5.59	8.41	9.71	9.34
Potential E-book reader customers planning to buy a device within 2 years	4.64	4.98	3.22	2.16

A. Pehlken et al. (Eds.), *Cascade Use in Technologies 2018*,
https://doi.org/10.1007/978-3-662-57886-5_6

TOTAL		10.23	13.39	12.93	11.5

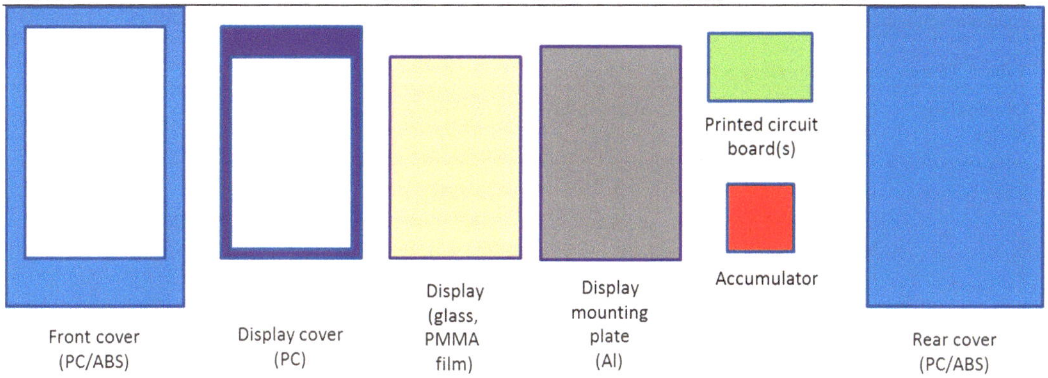

Figure 1. Main elements of an E-Book-Reader (schematic, including most important materials)

2 Materials and Methods

In total, ten discarded e-book readers were acquired for this study. Table 2 shows the models and makes used, which exhibit production dates between 2011 and 2015 and thus represent typical end of life products. Samples 4 to 6 were missing the batteries, and all of these devices showed a damaged rear cover. None of the devices was declared operational by the internet vendors.

Table 2. E-book reader samples used for experiments

Sample No.	Product/Model/ Producer	Year of production	Display size ['']	Circuit board board size [cm²]	Total mass [g]	Status (initially)
1	Kindle (7. Gen)/ WP63GW/ Amazon	2012	6.7	47.42	191.3	Damaged by water
2	Kindle (4. Gen)/ D01100/ Amazon	2011	6	44.72	167.6	Damaged by water
3	Kindle (4. Gen)/ D01100/ Amazon	2011	6	44.72	167.8	Damaged by water
4	hudl/HT7B1653/ Tesco	2014	7	70.56	281.2	Not operational*
5	hudl/HT7B1653/ Tesco	2014	7	70.56	280.7	Not operational*
6	hudl/HT7B1653/ Tesco	2014	7	70.56	285.9	Not operational*
7	nook/BNRV300/ B&N	2011	6	154.56	207.5	Damaged by water
8	Moshi Monsters/ MMU007D/ Ingo	2013	7	59.4	284.4	Does not boot
9	Paperwhite/EY2/ Amazon	2015	6.7	58.64	203.7	Damaged by water
10	Kurio 7S/C13000/ Kurio	2013	7	~	386.5	Display fissure

*accumulator was missing upon delivery ~device was not dismantled as hard reset operation was applied successfully

Nine of these devices were manually dismantled and the tenth device was reactivated by a reset. Dismantling was documented in detail identifying each single dismantling step along with the tools used and the time consumed. Firstly, a visual inspection was carried out, along with a functional test. This functional test used a "hard reset", i. e. a hardware reset by pressing the main switch for about 30 to 60 seconds. The dismantling tools afterwards used covered specific clip openers and opening tools for electronical devices ("Smartphone repair set"), tweezers, a bench vise, hammer, torx screwdriver 5x40, Philips screwdriver as well as a magnetic plate. A hot air gun Parlux 2800 (1760 Watts, air volume flow 67 m³/h) was used to remove glued connections. In the very right column of Table , the total number of tools used from this list was added to each of the samples. An infrared thermometer (RAY-Infrared Thermometer 31.1136, TFA Dostmann GmbH & Co. KG) was used for temperature measurements along with the hot air gun.

Dismantling documentation covered an assessment of dismantling difficulty (Table 3), dismantling time recordings, the generation of a dismantling graph and log and the generation of a bill of materials for each device.

Table 3. Dismantling difficulty rating scale

Dismantling Rating	Difficulty of manual operations	Accessibility	Level of destruction
1	No effort	easy	nondestructive
2	Little effort	medium	Partial destruction without impairing the function
3	Moderate effort	hard	destructive
4	Intensive effort	-	-

Besides the material marks of the (polymer) parts in the samples, nondestructive material identification was carried out as the basis for the bill of materials generation. A mobile XRF device (Niton TM XL2 RFA by Thermo Fisher) was applied for metals identification, using the XRF internal alloy identification standard [3].
The ATR-FTIR device (Alpha by Bruker), with its OPUS software and the Bruker polymer and filler database along with the S. T. Japan spectra library were used in order to identify the polymers [4].
To generate economic information, recyclability assessment (KE_M) according to VDI standard 2243 [5] was calculated, using recent LME data for metals, price quotes of a regional plastic waste management company for polymers, and a working wage of 0.26 Euro per minute. KE_M compares the disposal/production with the recycling costs of a single part, and with a value higher than "1" indicates that recycling might be economical.

3 Results

The main information acquired from the dismantling trials was a complete bill of materials, which is shown in Table 4. Shares of single materials have been calculated of the single samples by calculating the arithmetic mean value. Results show that the most important materials are glass (19.8%), printed circuit boards (14.3%), and the housing materials (amongst others, PC-GF with 10.4%). The share of accumulator mass of all the samples ranged from 9.7 % (sample 1) to 20.6 % (sample 8). The highest single mass identified in the dismantling steps was the zinc (59.8-61.7 g) or aluminum components (17.1-54.9 g), which served as mounting base (plate) for the display.

Table 4. Bill of material of samples 1 to 9

Material [g] / Sample %	1	2	3	4	5	6	7	8	9	TOTAL [g] / %
PC/GF	-	-	-	69.1	71.3	69.1	-	-	-	209.5 / 10.4%
PC/ABS	63.2	-	-	9.1	9.2	9.2	55.5	-	-	146.2 / 7.3%
PC	9.5	38.1	38.1	15.5	13.4	14.2	-	12.2	52.6	193.6 / 9.6%
PMMA	12.9	-	-	5.2	5.2	5.6	-	71.0	-	99.9 / 5.0%
PET	-	5.1	5.1	11.8	11.0	11.9	8.2	11.1	-	64.2 / 3.2%
Zn	-	-	-	60.4	61.7	59.8	-	10.0	-	191.9 / 9.5%
Al	17.1	27.1	27.2	-	-	-	54.9	-	18.4	144.7 / 7.2%
Fe	-	9.5	9.3	-	-	-	-	30.0	-	48.8 / 2.4%
Glass	39.1	38.6	38.4	59.6	59.2	59.8	18.0	43.6	42.0	398.3 / 19.8%
Printed Circuit Board	27.4	22.7	22.7	41.2	40.6	40.2	32.5	32.8	26.9	287.0 / 14.3%
accumulator	18.2	18.3	18.3	n/a	n/a	n/a	31.4	55.7	33.8	175.7 / 8.7%
Other	0.6	8.1	8.6	6.8	6.5	6.6	6.5	4.2	2.3	50.2 / 2.5%
TOTAL 100%	188.0	167.5	167.7	278.7	278.1	276.4	207.0	270.6	176.0	2010.0 /
Accumulator share [%]	9.7%	10.9%	10.9%	n/a	n/a	n/a	15.2%	20.6%	19.2%	

Figure 2 shows the medium composition of the samples 1 to 3 and 7 to 9 which were the samples delivered and surveyed completely, i. e. including the accumulator. Samples 4 to 6 results were not included in Figure 2 as they were lacking their batteries. As part of the pre-treatment requirements of the ElektroG, it is pointed out that initial treatment must be carried out prior to the implementation of recovery or disposal measures. At least all liquids as well as the substances, mixtures or components which are listed in Annex 2 of the ElektroG (selective treatment of materials and components of old appliances) have to be removed [2]. Relevant components for this work are the accumulators and the circuit boards (> 10 cm²). Consequently, during the experiments the effort for removing accumulators and larger printed circuit boards was paid attention to.

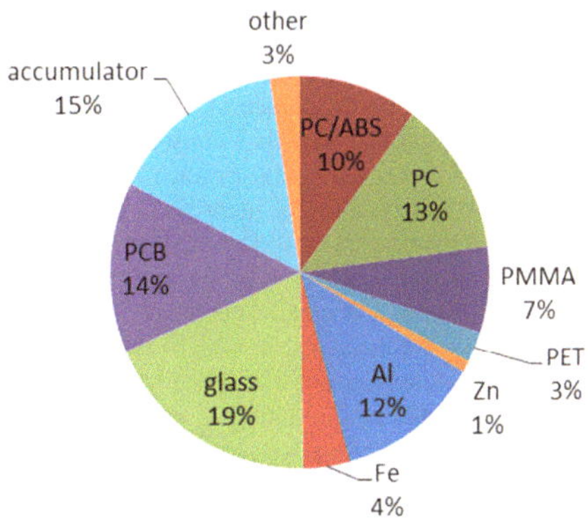

Figure 2. E-book reader mean composition (in mass percent) of samples 1-3 and 7-9

Table 5 shows the results of the dismantling time in sum values, i.e. the total duration from the very start of the dismantling process to the removal of the accumulator, the printed circuit board, the display and the last module of the entire e-book reader, respectively. Regarding the arithmetic mean values of these dismantling times, accumulator dismantling requires about 110 seconds of total dismantling time, the printed circuit boards removal consumes 155 seconds from the start, display removal takes 445 seconds, and the mean total dismantling time was found to be about 592 seconds if the dismantling method was followed which is targeted at a parts re-use, i.e. trying to work in a non-destructive way.

Table 5. Characteristic dismantling times (total sum from beginning of dismantling) for accumulators, printed circuit boards, display, and complete dismantling

Sample No.	Dismantling time for	Accumulator [s]	Printed Circuit Board [s]	Display [s]	Total dis-mantling [s]
1		212	268	926	926
2		103	338	877	877
3		86	341	860	860
4		n/a	19	236	451
5		n/a	17	225	546
6		n/a	30	245	481
7		63	111	217	477
8		96	146	228	237
9		98	123	192	474
Arithmetic mean value		109.7*	154.8	445.1	592.1
n/a = not available		*based on the sample 1-3 and 7-9 results only			

Table 6. Fastening characteristics of the individual samples

Sample No.	No. of tools used [#]	No. of screws [#]	No. of clips [#]	Adhesive area [cm²]	Adhesive connection of accum.	Display	cover (f)
1	8	27	24	211.9	y	y	y
2	7	15	26	179.15	y	y	y
3	7	15	26	179.15	y	y	y
4	4	5	25	95,65	n/a	y	n
5	4	3	25	95.2	n/a	y	n
6	4	3	25	96.45	n/a	y	n
7	6	11	32	76.76	y	y	y
8	4	9	44	77.22	y	y	n
9	6	19	15	149.94	n	y	y
Arithmetic mean value	5.6	11.9	26.9	129.05			
#= number	accum. = accumulator	cover (f) = front cover		n/a = not available		y = yes	n = no

Dismantling time is mainly determined by the joining technique chosen, along with visibility and accessibility of the connections. Mainly screws, clip connections and adhesive bonding were the connections which were found in the samples. But even if identical connecting techniques are applied, due to individual characteristics of the samples, dismantling times might differ remarkably: Taking sample 2 and sample 3 as an example (both are Kindle 4th generation products), a time difference of about 63 seconds for removing the accumulator was identified (sample 2: dismantling time of 83 seconds, sample 3 consumed only 20 seconds for this single dismantling step).

Thus in Table 6 the characteristics of the connections of each individual sample were collected. The total amount of different tools used was identified, along with the number of screws and clips to be opened. Moreover, the total adhesive bonds area was identified, using a ruler to measure the adhesive surface area, the adhesive surfaces visible and adding up single areas. The devices with the glued display units can be identified using this comparison. Moreover, sample 7 shows the smallest adhesive surface area, because the connection mechanism used here was an approximately 0.7 cm wide adhesive strip which was placed along the edge of the display. As a result, the arithmetic mean adhesive area of the samples was 129.05 cm^2.

Moreover, due to the adhesive bonds between the display unit and the mounting plate only destructive dismantling using increased effort was possible. Because of the variations in each dismantling step, the dismantling rating scale (cf.Table 3) was used to assess the dismantling effort on a sample specific scale, and to mathematically identify a mean dismantling rating based on an expert's rating. To this end, the arithmetic mean value can be calculated as the sum of a criterion (e.g. sum of difficulty sample 1 = 23) divided by the number of dismantling steps (number of dismantling steps sample 1 = 12). Thus a more robust statement can be made on the overall rating of the ease of dismantling. Table 7 shows the evaluation criteria for the summation of all dismantling steps of the samples. The easiest dismantling process can be found with sample 8, whereas accessibility was best with sample 3 and minimal destruction was observed with sample 1. Comparison of different model years of one make can be made between sample 1 (Kindle 2012) and samples 2 and 3 (Kindle 2011), showing basically little change except for fewer dismantling steps.

Table 7. Dismantling assessment of the individual samples

Sample No.	Total dismantling [s] A	Total dismantling steps [#] B	Average time [s] B/A	Avg. dismantlg. difficulty (max. value) [-]	Avg. accessibility (max. value) [-]	Avg. destruction level (max. value) [-]
1	926	12	76	1.92 (4)	1.58 (3)	1.17 (3)
2	877	15	59	1.93 (4)	1.47 (3)	1.20 (3)
3	745	15	50	1.60 (4)	1.13 (3)	1.20 (3)
4	451	9	50	2.11 (4)	1.78 (3)	1.56 (3)
5	546	9	61	2.11 (4)	1.78 (3)	1.56 (3)
6	481	9	53	2.00 (4)	1.89 (3)	1.78 (3)
7	477	9	53	2.00 (4)	1.44 (3)	1.22 (3)
8	237	19	18	1.60 (3)	1.30 (2)	1.20 (3)
9	474	8	61	1.88 (4)	1.63 (3)	1.25 (3)
Arithmetic mean value		10.7	53.4	1.91	1.56	1.35

Avg. = average

Dismantling time data collection formed the basis of the recyclability assessment by calculating the KE_M parameter according to the VDI 2243 standard. Calculation was carried out for each of the materials identified in Table 4. Results showed values exceeding the threshold of "1" only for the zinc parts in samples 4 (KE_M=1.099), 5 (KE_M=1.124) and 6 (KE_M=1.179), but with a very low margin for each. Thus under the given dismantling conditions economical operations seem, at the very least, unlikely.

Dismantling improvements may be seen from automated dismantling operations such as introduced by Apple ("Daisy" robot, [6]) for smartphones for future developments. Future research is needed to quantify the printed circuits composition as a basis for economical assessment. Earlier research [7] identified e. g. gold bonding wires and silver parts on e-book reader circuit boards, but their quantification and a comparison with smartphone circuit boards concentrations is still to be carried out.

4 Conclusion

The arithmetic mean mass of all e-book readers surveyed in this study was 207.1 g, including three products acquired without an accumulator. The arithmetic mean mass of the devices which were acquired in a complete state (samples 1-3 and 6-9) was 203.7 g before dismantling.

The mean e-book reader composition found was 30 % polymers (PC, PC/ABS, PMMA, PET), 17 % metals (Zn, Al, Fe), 19 % glass, 14 % printed circuit boards, and 15 % accumulator, and 3 % of other materials such as cables and

connectors. Besides the afore-mentioned fractions, the largest single materials share was PC (13%) and aluminum (12%).
Identification of the cumulative dismantling times of individual parts yielded arithmetic mean values of about 110 seconds for the accumulator, 155 seconds for the printed circuit board, and 445 seconds for the dismantling of the display, with a mean total dismantling time of 592 seconds. This dismantling time duration renders recyclability for almost all materials identified as uneconomical, except for some zinc parts. This fact can be attributed to the high number of connections which needed to be opened (arithmetic mean values were 11.9 screws and 26.9 clips to be opened, using 5.6 different tools for these operations for one device). Typical dismantling characteristics were 10.7 dismantling steps per device, with an arithmetic mean duration of 53.4 seconds per dismantling step. Thus from a purely economical perspective, mechanical recycling should be preferred over manual dismantling for recycling of e-book readers.

5 Zusammenfassung

Derzeit werden in Deutschland etwa 10 Millionen E-book-Reader genutzt. Aufgrund der abnehmenden Marktdurchdringung wurde zur Bestimmung des Recyclingpotentials dieser Geräteklasse eine Demontageanalyse durchgeführt. Insgesamt wurden zehn defekte Geräte von fünf verschiedenen Herstellern beschafft, von denen neun manuell demontiert und der zehnte nach dem Zurücksetzen auf Werkseinstellungen („hard reset") wieder in Betrieb genommen werden konnte. Bei der Demontageuntersuchung wurden der Zeitbedarf der Zerlegeschritte, der Demontageaufwand, die Zugänglichkeit der Verbindungen und die Möglichkeit der zerstörungsfreien Demontage erfasst und bewertet.
Die Hauptkomponenten von E-book-Readern werden von den beiden Gehäusehälften (Vorder-/Rückseite), dem Displaymodul (Display, Montageplatte, Displayabdeckung), Leiterplatten und dem Akkumulator gebildet. Der arithmetische Mittelwert der Massen der Untersuchungsobjekte lag bei 207,1 g pro Gerät. Der entsprechende Wert nur für die Geräte, die vollständig einschließlich Akkumulator angeliefert wurden (Geräte 1 bis 3 und 6 bis 9), lag bei 203.7 g. Zur Bewertung der Demontierbarkeit wurden drei Parameter mit einer drei- bzw. vierstufigen Skala bei der Demontage bewertet. Ergänzend wurden Röntgenfluoreszenz- und ATR-FTIR-Untersuchungen zur Materialidentifikation durchgeführt.
Im Ergebnis wurde eine mittlere Zusammensetzung (arithmetischer Mittelwert) von 30 % Kunststoffen (PC, PC/ABS, PMMA, PET), 17 % Metalle (Zn, Al, Fe), 19 % Glas, 14 % Leiterplatten sowie 15 % Batterie und 3 % sonstiger Kompoenten (Kabel, Verbindungselemente) gefunden. Innerhalb dieser vorgenannten Werkstoffgruppen sind wichtige Einzelfraktionen PC (13%) und Aluminium (12%). Die Erfassung der kumulativen Demontagezeit einzelner relevanter Bauteile zeigte mittlere Demontagezeiten für die Batterien von etwa 110 Sekunden, für die Leiterplatten von 155 Sekunden und von 445 Sekunden für das Display, bei einer mittleren Gesamt-Demontagezeit von 592 Sekunden. Durch diese hohen Demontagezeiten ist eine manuelle Demontage zum Werkstoffrecycling praktisch nicht ökonomisch durchführbar, was auch anhand der Bestimmung der KE_M-Werte nach VDI 2243 für die gewonnenen Werkstoffe gezeigt werden konnte. Der Grund hierfür liegt unter anderem in der hohen Anzahl der zu lösenden Verbindungselemente (im Mittel 11,9 Schrauben und 26,9 Clip-Verbindungen), die im Schnitt die Nutzung 5,6 verschiedener Werkzeuge erfordern. So ergeben sich im Mittel 10,7 Demontageschritte pro Gerät mit einer mittleren Länge von 53,4 Sekunden pro Schritt. Im Ergebnis muss für diese Geräte eine mechanische Aufbereitung zur Werkstoffrückgewinnung empfohlen werden.

6 Acknowledgements

The authors would like to thank Mrs. Gabriella Loveday of Hochschule Pforzheim (Germany) and Dr. Thorsten Hartfeil of Fraunhofer ISC Projektgruppe für Wertstoffkreisläufe und Ressourcenstrategie IWKS in Alzenau (Germany) for their support.

7 References

[1] IfD Allensbach. "Anzahl der momentanen und potenziellen Nutzer von E-Readern in Deutschland von 2013 bis 2016 (in Millionen)." Statista - Das Statistik-Portal, Statista, de.statista.com/statistik/daten/studie/170040/umfrage/einstellung-zur-nutzung-von-e-books/, Accessed 6. April 2018

[3] Schwedt, Georg: Analytische Chemie. Grundlagen, Methoden und Praxis, Atomspektroskopische Methoden: Röntgenfluoreszenzanalyse, 2„ vollständig überarbeitete Auflage. WILEY-VCH Verlag, Weinheim, 2008

[4] Hesse, Manfred; Meier, Herbert; Zeeh, Bernd: Spektroskopische Methoden in der organischen Chemie, Infrarot- und Raman-Spektren: IR-Spektrometer, Georg Thieme Verlag, Stuttgart, 2012

[8] Schischke, Karsten; Stobbe, Lutz; Scheiber, Sascha; Oerter, Markus; Nowak, Torsten; Schlösser, Alexander; Riedel, Hannes; Nissen, Nils: Disassembly Analysis of Slates: Design for Repair and Recycling Evaluation, Fraunhofer IZM, Berlin, 2013

[2] Elektro- und Elektronikgerätegesetz (ElektroG) Gesetz über das Inverkehrbringen, die Rücknahme und die umweltverträgliche Entsorgung von Elektro- und Elektronikgeräten, Bundesgesetzblatt 1 S. 1739, 2015

[5] Verein Deutscher Ingenieure (VDI, Hrsg.): VDI 2243 Recyclingorientierte Produktentwicklung, Recyclingorientierte Gestaltung. VDI-Verlag, Düsseldorf, 20020
[6] https://www.apple.com/environment/resources/, checked on June 2, 2018
[7] Ott, Laura: Potentialanalyse der Rückgewinnung von Sekundärrohstoffen bei E-Book-Readern. Bachelor Thesis. Pforzheim University, Pforzheim, 2016

Extraction Potential of Tantalum from Spent Capacitors Through Bioleaching

Mehmet Ali Kucuker[1], Xiaochen Xu[1] and Kerstin Kuchta[1]
[1] Institute of Environmental Technology and Energy Economics, Hamburg University of Technology, Hamburg, 21079, Germany
kucuker@tuhh.de

Abstract

Tantalum (Ta) is the one of the most critical elements according to the European Commission. There is limited research on tantalum recovery from secondary sources such as waste electrical and electronic equipment (WEEE), bottom ash and by products of the industrial activities. In this study, the recovery potential of tantalum from spent tantalum capacitors was tested using bioleaching under the biomining concept. Three different kinds of microorganisms were tested for tantalum recovery, which were *Pseudomonas putida* (DSM No. 6125), *Bacillus subtilis* (DSM No. 1088532), and *Penicillium simplicissimum* (DSM No. 1078). It turned out that *P. simplicissimum* has the ability to leach tantalum from wasted tantalum capacitors. The maximum leaching rate was 1.25 g Ta per kg sample, 0.67% of Ta could be extracted after a period of 14 days. An unknown species achieved the highest leaching rate (9.88 g Ta / kg sample, extraction rate of 5.31%) for 15 days, at 25 °C and 150 rpm, bulk density of 0.1%, but the isolation and identification failed. The potential of tantalum recovery by bioleaching is demonstrated, however, further research needs to be carried out.

1 Introduction

Tantalum is a hard, blue-gray transition metal with high corrosion-resistance. Tantalum is highly conductive of heat and electricity, and is easily fabricated, in addition to having a very low coefficient of thermal expansion [1]. These features of tantalum result in a very wide application. It can be used in the instruments for preparation of inorganic acids, and its service life could be ten times longer than that of stainless steel. In chemical, electronics and electrical industries, tantalum can also be used to take the place of some precious metals, and thus reduce the cost. Tantalum is also a kind of biological metal. It can be used as orthopedic and surgical material, and silk, to reconnect broken nerves. Tantalum carbide is very hard, and has a high melting point (3880°C), so it can be used for cutting tools and drills. Tantalum oxide is used for the manufacturing of advanced optical glass and catalysts [1].

More than half of the total tantalum is used for tantalum capacitors. Tantalum capacitors are widely used in communication equipments, digital audio and video products, computers, automotive electronics and the defense industry. The characteristics of tantalum capacitors are attractive. Ueberschaar et al. [2] reported the average tantalum content per mass printed circuit board (PCB) and product (Figure 1). In addition, elemental composition of tantalum capacitors were identified by Ueberschaar et al. [2] (Figure 2). Compared with traditional aluminum capacitors, tantalum capacitors are more stable. The leakage current is much lower, as is capacitor heating, which might have some influence on the service life. In addition, the failure rate of tantalum capacitors is also lower [3, 4].

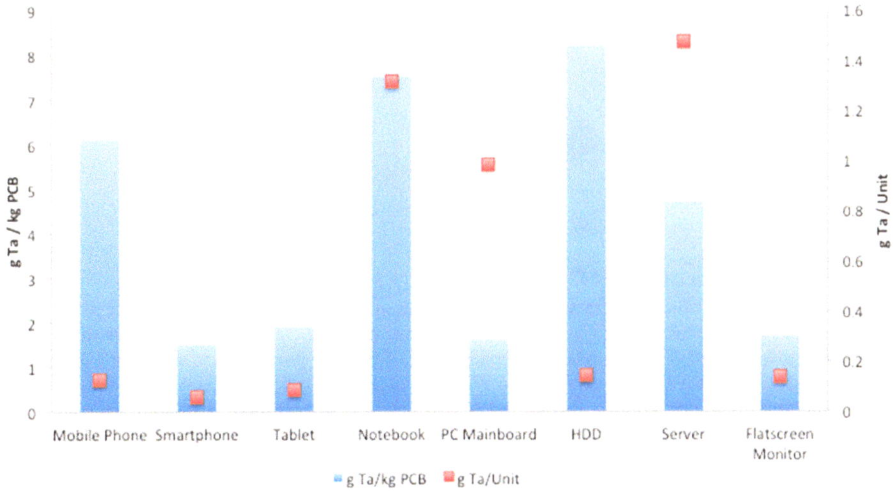

Figure 1. Average Ta content of the PCB and product (adapted from [2])

A. Pehlken et al. (Eds.), *Cascade Use in Technologies 2018*,
https://doi.org/10.1007/978-3-662-57886-5_7

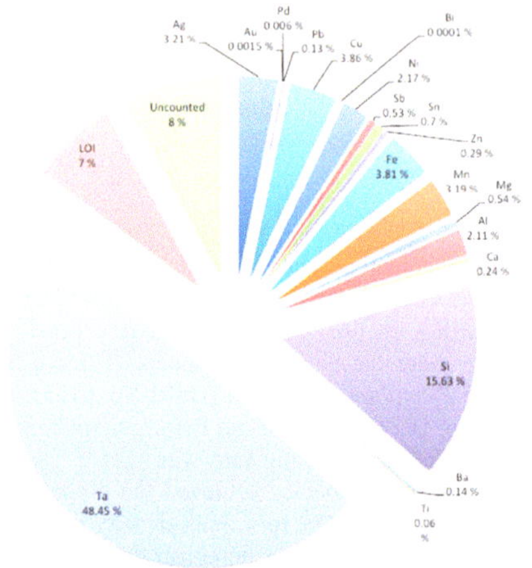

Figure 2. Elemental composition of tantalum capacitors (LOI: Loss of Ignition) (adapted from [2])

1.1 Reserves and production

According to United States Geological Survey (USGS), the global reserves of tantalum are larger than 100,000 tons, of which about 62% are found in Australia and about 36% in Brazil [5]. The largest two tantalum deposits are both in Australia. In 2014, the global tantalum mineral production was 1200 tons, of which the production from Rwanda accounts for half of the total production (600 tons). Figure 3 shows the global tantalum mineral production in 2014 [5].

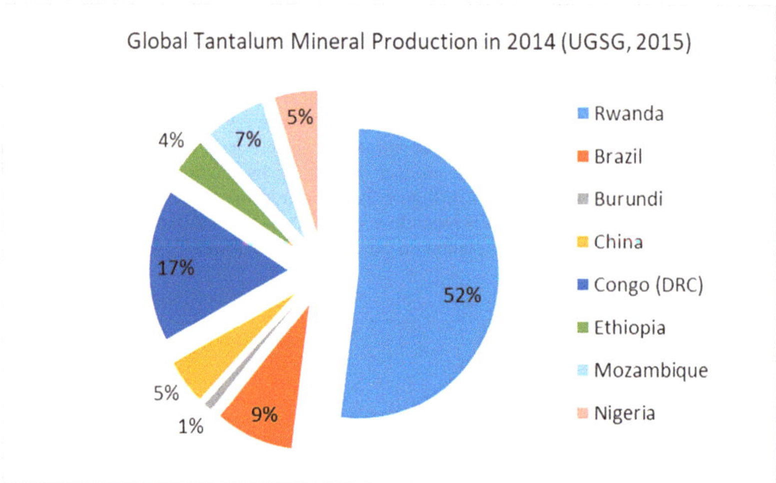

Figure 3. Global tantalum production in 2014 [5]

1.2 Recovery of Tantalum from WEEE

In WEEE or e-waste, most of the valuable substances are found in PCBs [6]. Precious and scarce materials account for only a small percentage of the total weight, but with enough recycling value. Tantalum accounts for about 0.0157% of the total weight of a computer and is mostly found in capacitors [7]. Since the production of tantalum has not been stable in price and the quantity, the recovery of tantalum from WEEE (mainly used capacitors) is becoming a serious issue. There is a general trend towards miniaturization and the use of less tantalum per capacitor; therefore, the quantity of tantalum powder per tantalum wire has been decreasing [2]. As a result of this situation the implementation of a recycling process for tantalum from waste capacitors may not be sustainable [2]. Nowadays, about 15% to 20% of the total tantalum production is from waste recycling processes [8]. The normal tantalum recovery process is quite complex, especially for the metal cladding liquid tantalum capacitors. Electrolysis or aqua regia degradation has to be applied first to remove the metal clad, followed by sodium reduction or carbon reduction for deoxidation, and electron beam smelting is applied as the last step (Figure 4-A). In addition, for chip capacitors, after the separation process which removes plastic and wires, sodium hydroxide, nitric acid and hydrochloric acid should be applied step by step to leach and remove tin, silver and manganese, then sodium reduction and electron beam smelting are applied (Figure 4-B) [8-10].

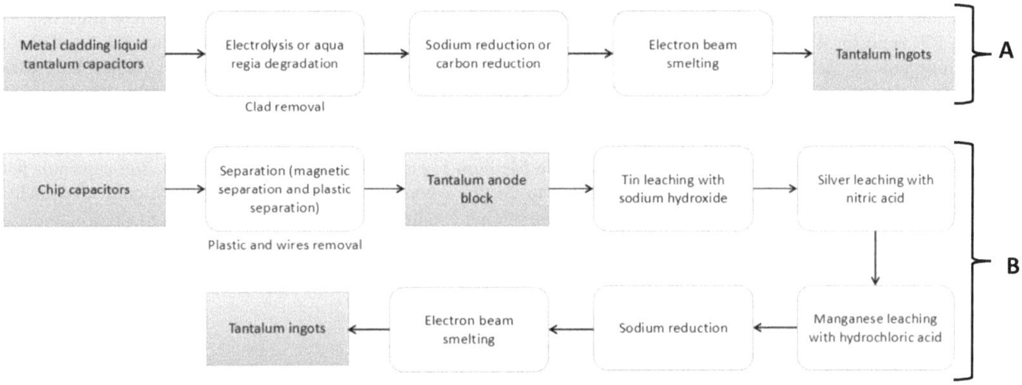

Figure 4. Tantalum recovery process for the metal cladding liquid tantalum capacitors (A), for chip capacitors (B) [8, 9]

1.3 Aims of the study

As mentioned before, bioleaching processes are already widely used in the mining industry. A lot of designs and innovative ideas are still at lab scale. There is only limited research on the recovery of precious metals (PMs) and rare earth elements (REEs) from WEEE using biomining process. But rough ideas are already given. For example, Gurung et al. [11] carried out gold and silver recovery through biomining from spent mobile phones, and achieved a maximum recovery rate of 72.33% for gold and 85.91% for silver. Kucuker [6] reported that PMs and REEs could be recovered from WEEE using biomining process such as bioleaching and biosorption. So the possibility of applying biomining to the recovery of tantalum, which is more or less alike to precious metals and rare earth elements, has already been shown. The aim of this study is to test the bioleaching method for the recovery of tantalum from spent capacitors. Three different microorganisms (*Pseudomonas putida* (DSM No. 6125), *Bacillus subtilis* (DSM No. 1088532), and *Penicillium simplicissimum* (DSM No. 1078)) were studied.

2 Methods and Materials

2.1 Characterization of the Sample of Spent Tantalum Capacitors

The tantalum capacitor samples were obtained from WEEE and were milled into powder. A special acid digestion with hydrofluoric acid (HF) was carried out in order to measure the tantalum content in the sample using the EU standard method EN 13656 (Characterization of waste - Microwave assisted digestion with hydrofluoric (HF), nitric (HNO_3), and hydrochloric (HCl) acid mixture for subsequent determination of elements) [9]. 0.5 g of milled tantalum capacitor samples were taken, and filled into each plastic digestion vessel. The samples were pre-humidified with a few drops of deionized water. An acid mixture of 6 mL HCl, 2 mL HNO_3 and 2 mL HF was filled into each digestion vessel. In the next step, the digestion vessels were closed, put into the microwave oven and the digestion program was applied according to the DIN EN 13656 method.

The microwave-assisted digestion was finished, the samples were cooled and the digestion vessels were opened. To avoid potential harm to the analysis devices, the hydrofluoric acid was neutralized by building up a complex with boric acid. 650 mg boric acid was added into each vessel (e.g. 22 ml 4 % (m/m) solution). The vessels were closed and heated again in the microwave at 300 W for 3 min.

Afterwards the vessels were cooled down. To deal with the particles remaining in the vessels, the solution could be either centrifuged or filtrated. The solution was first analyzed qualitatively with energy dispersive X-ray fluorescence (EDXRF) and then quantitatively analyzed with Inductively Coupled Plasma Optical Emission Spectrometer (ICP-OES).

2.2 Bioleaching Experiments

Three different microorganisms were chosen for the bioleaching experiment: *Pseudomonas putida* (DSM No. 6125), *Bacillus subtilis* (DSM No. 1088), and *Penicillium simplicissimum* (DSM No. 1078). All three species were ordered from Deutsche Sammlung von Mikroorganismen und Zellkulturen GmbH (DSMZ) [12]. Two mediums (Medium 1 and Medium 2) were prepared for the cultivation of microorganisms and the bioleaching experiment. Medium 1: 5.0 g peptone and 3.0 g meat extracts were added into 1000 ml distilled water, and the pH value was adjusted to 7.0. Medium 2: 200 g scrubbed and sliced potatoes were boiled in 1000ml water for 1 hour. The mixture was then sieved, and 20 g glucose was added into every 1000ml sieved infusion. Both mediums were autoclaved before further operations. *Pseudomonas putida* (DSM No. 6125) and *Bacillus subtilis* (DSM No. 1088) were cultivated in Medium 1 at 28 °C, 150 rpm, and *Penicillium simplicissimum* (DSM No. 1078) was cultivated in Medium 2 under room temperature (about 20-23 °C), 150 rpm, for 48 hours. A test was carried out to check the potential of bioleaching of tantalum with the chosen microorganisms. Before the test, a literature survey was carried out, aimed at figuring out suitable bioleaching conditions for the test experiment. Two-step bioleaching was chosen as a test method. Tantalum capacitors were grinded into particles with a size of about 0.5 mm. 100 ml of each culture were poured into 250 ml flasks, and 0.1 g tantalum capacitor particles were added into each flask, to achieve a bulk density of 0.1%. Every test was duplicated, and 3 comparative tests were carried out, with 100 ml deionized water, pure Medium 1, pure Medium 2, and 0.1 g tantalum capacitor particles. The mouths of the flasks were covered by cotton to ensure a good aeration. All 9 flasks were cultivated under room temperature (about 20-23 °C) and 150 rpm. To ensure mixing conditions, all the flasks were shaken slightly by hand for 1 min. 5 ml samples were taken from each flask on the 7th and 15th day for the measurement of tantalum ion concentration.

3 Results and Discussion

3.1 Characterization of the Tantalaum Capacitors

In Figure 5, the qualitative analysis result of the tantalum capacitor sample by the EDXRF analysis is shown. From the spectrum, Ta, Ti, Ba, Fe, Mn, Nr, W, Pb, BR, Zr, Nb and Ag (out of the diagram) can be detected from the sample. The Tantalum content measurement was carried out in the central laboratory for chemistry analytics in the Hamburg University of Technology (TUHH), following the standard method DIN EN 13656. The Ta content in the representative sample was 185.9 g Ta / kg sample.

Figure 5. The elemental composition of the Ta-capacitor sample by EDXRF

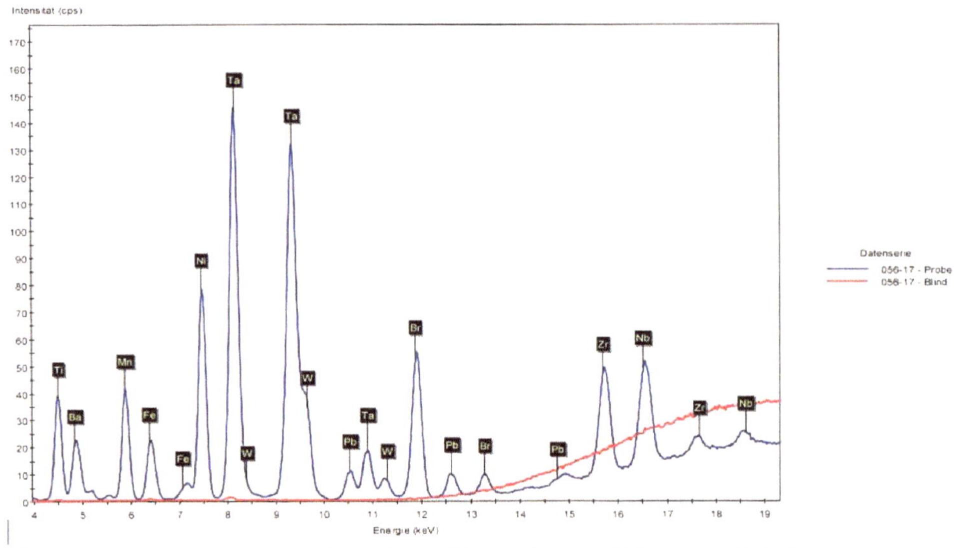

According to the literature, the content of tantalum in Ta capacitors could be around 40%. But from the spent Ta capacitors the characterization showed that the content of tantalum was 18.59%. Instead, the wasted Ta capacitors contained quite high percentages of Fe, Cu, Ni, and Pb. Since the capacitors were sorted from wasted PCBs, it is certain that the collected capacitors contained a certain amount of wires and impurities. As the capacitors are small and hard to be recognized, some other types of capacitors, such as Ni capacitors and Al capacitors might be mixed together with Ta capacitors. This can explain the low content of Ta and high content of other metals in the wasted Ta capacitors.

3.2 Bioleaching Tests

The results of the bioleaching experiments are given in Table 1. In the bioleaching experiments, tantalum essentially cannot be leached out by *Pseudomonas putida* (DSM No. 6125, maximum leaching rate: 0.33 g Ta / kg sample, Ta extraction rate of 0.18%) and *Bacillus subtilis* (DSM No. 1088, maximum leaching rate: 0.34 g Ta / kg sample, Ta extraction rate of 0.18%) even under favorable growth conditions. *Penicillium simplicissimum* (DSM No. 1078) has the ability to leach tantalum, but the leaching rate was still limited (maximum leaching rate: 1.25 g Ta/ kg sample, extraction rate of 0.67%). An unknown species achieved the highest leaching rate (9.88 g Ta / kg sample, Ta extraction rate of 5.31% in 15 days, and 28.65 g Ta / kg sample, Ta extraction rate of 15.41% after 70 days), at 25 °C and 150 rpm, bulk density 0.1%, but the isolation and identification of this species failed. The unknown species which contaminated the flasks in the prior experiments achieved the highest leaching rate. As reference group Medium 2 was the contaminated group, the unknown species could possibly be a kind of fungi. The problem is the isolation of the unknown species failed, which means the microorganisms in the contaminated flasks were already dead after the cultivation period for bioleaching. The contamination was successfully repeated for one more time. However, this is an endless loop, because by the time the leaching result turns out that the contamination occurred by the right species, the culture is already dead. This leads to a very high identification cost for the species, as metagenome sequencing plus the data analysis have to be done to identify the species.

As the content of tantalum in PCBs was extremely low, the exact percentage-leaching rate cannot be calculated. With regards to the low content of tantalum in PCBs, it is not commercially acceptable to recover tantalum as a single metal directly from PCBs without a combination with other metal recovery processes, but as the bioleaching process also has the ability to leach out a high percentage of base metals (copper, nickel, aluminum and manganese), the combined multi-metal recovery process can be taken into account. Tantalum is a very stable metal element which cannot be leached even by aqua regia, and there is almost no current existing literature on tantalum bioleaching. As the pioneer of this kind of research, this study started from zero. The potential of bioleaching is clearly shown. As the bioleaching process does not require and produce highly toxic and highly corrosive chemicals, the downstream processes, including metal removing from liquid phase and waste water treatment, would also be much simpler. The research in tantalum recovery by bioleaching is interdisciplinary, and needs very close cooperation between different labs and institutes, especially for unexpected situations, as in this case, for example, the contamination of the reference group. The identification of unknown microorganisms (Denaturing Gradient Gel Electrophoresis, DGGE, for example) needs special equipment and expertise, so it took time and labor to contact and cooperate with another institute (Technical Microbiology, TUHH). In addition, funding is needed for further identification processes.

Table 1. Results of the bioleaching experiments

Bulk density 0.1%	**Leaching rate of Ta (%)**		**Bulk density 1%**	**Leaching rate of Ta (%)**	
Leaching agent	**7th day**	**15th day**	**Leaching agent**	**7th day**	**15th day**
Medium 1	ND	0.091	Medium 1	0.002	0.019
Medium 129	2.495	5.314	Medium 129	0.168	0.389
P. simplicissimum 1	0.295	0.672	*P. simplicissimum 1*	0.123	0.293
P. simplicissimum 2	0.398	0.591	P. simplicissimum 2	0.115	0.249
P. putida 1	0.155	0.172	*P. putida 1*	0.006	0.021
P. putida 2	ND	0.177	*P. putida 2*	ND	0.026
B. subtilis 1	0.016	0.102	*B. subtilis 1*	ND	0.017
B. subtilis 2	0.166	0.182	*B. subtilis 2*	ND	0.037

ND: under detection limit

For bioleaching experiments, prior research shows that smaller particle size of raw materials, low bulk density and better growth conditions may have the potential to achieve higher leaching rates. It is also recommended to test more different microorganism types and species, and different leaching processes (for example, spent medium leaching). Current existing literature also figured out the possibility to use mixed cultures (with different microorganisms) in bioleaching. This is already applied in gold mining, but for tantalum it still needs more effort. Genetic modification might also be helpful to increase the leaching rate. The research done by TUHH has certainly figured out the potential of tantalum recovery by bioleaching, as the research is still in its infancy.

4 Conclusion

The result of the tantalum recovery trial by bioleaching from tantalum capacitors turns out to be negative. Tantalum essentially cannot be leached out by *Pseudomonas putida* (DSM No. 6125, maximum leaching rate: 0.33 g Ta / kg sample, extraction rate of 0.18%) and *Bacillus subtilis* (DSM No. 1088, maximum leaching rate: 0.34 g Ta / kg sample, extraction rate of 0.18%) even under favorable growth conditions. *Penicillium simplicissimum* (DSM No. 1078) has the ability to leach tantalum, but the leaching rate was still limited (maximum leaching rate: 1.25 g Ta / kg sample, extraction rate of, 0.67%). An unknown species achieved the highest leaching rate (9.88 g Ta / kg sample, extraction rate of 5.31%), at 25 °C and 150 rpm, bulk density 0.1%, but the isolation and identification failed. Since there is no current existing literature that has directivity in the field of tantalum bioleaching, it is difficult to choose the species for experiment, as there are different kinds of microorganisms which can produce acid and have the potential for bioleaching. This means it might be constly to find out a suitable species. However, researchers focus on species which might be helpful while studying the mechanism of tantalum bioleaching. On the other hand, genetically modified organisms might be the best choice to increase the bioleaching rate of tantalum. However, this only feasible after the mechanism of tantalum leaching is totally clarified. Tantalum cannot be leached out even by aqua regia, so a high leaching rate must be caused by the synergy of several different organic or inorganic substances. It might be quite hard to figure out the actual mechanism, and the genetic modifying even after the mechanism is clarified, since the research will take a huge amount of time, and needs cross-sectoral cooperation.

5 Zusammenfassung

Tantal wird von der Europäischen Kommission als hoch kritisches Element eingestuft, dennoch existieren bisher kaum Forschungsarbeiten zur Rückgewinnung von Tantal aus Sekundärquellen wie Elektro- und Elektronikaltgeräten, Schlacke und Nebenprodukten industrieller Aktivitäten. In diesem Beitrag wird das Rückgewinnungspotenzial von Tantal aus verbrauchten Tantalkondensatoren durch Bioleaching untersucht. Hierfür wurden die drei verschiedenen Mikroorganismen *Pseudomonas putida* (DSM No. 6125), *Bacillus subtilis* (DSM No. 1088532), und *Penicillium simplicissimum* (DSM No. 1078) getestet. Es stellte sich heraus, dass *P. simplicissimum* Tantal aus gebrauchten Tantalkondensatoren auslaugen konnte. Die höchste Auslaugungsrate betrug hierbei 1,25g Tantal pro kg Versuchsmaterial, und nach einer Periode von 14 Tagen konnten 0,67% Tantal extrahiert werden. Eine unbekannte Spezies erreichte die höchste Auslaugungsrate (9,88 g Ta / kg Versuchsmaterial, Extraktionsrate von 5,31%) bei 15 Tagen, 25°C, 150 rpm und einer Lagerungsdichte von 0,1%, jedoch sind Isolation und Identifizierung fehlgeschlagen. Das Potenzial der Tantalrückgewinnung durch Bioleaching konnte gezeigt werden, es sind jedoch weitere Forschungen notwendig.

Acknowledgments. The authors acknowledge the financial support from BMBF (The Federal Ministry of Education and Research) through the project [a]Biotechnological Approach For Recovery Of Rare Earth Elements And Precious Metals From E-Waste (BIOREEs), Project Number: 01DL14004. We are grateful to Asma Sikander, Ayah Alassali, Julia Hobohm, Nils Wieczorek, and Jinyang Guo as they helped us during the experimental study.

6 References

[1] Wikipedia, Tantalum Search, https://en.wikipedia.org/wiki/Tantalum, access18/3/2017.

[2] Ueberschaar M, Dariusch Jalalpoor D, Korf N, Rotter VS (2017). Potentials and barriers for tantalum recovery from waste electric and electronic equipment. Journal of Industrial Ecology, 21(3): pp 700-714.

[3]Information of tantalum capacitors, http://www.jb51.net/hardware/MotherBoard/13078.html, access 30/3/2017.

[4] Zhong J (2003). Research status and development trend of chip tantalum capacitors. Rare Metals. 11, pp 1-3.

[5] Tantalum Resources Reserves and Mineral Production, http://baike.asianmetal.cn/metal/ta/resources&production.shtml, access 19/3/2017.

[6] Kucuker M (2018). Biomining Concept for Recovery of Rare Earth Elements (REEs) from Secondary Sources. Verlag Abfall aktuell der Ingenieurgruppe RUK GmbH, Stuttgart, Hamburger Berichte; Bd. 48.

[7] Wencheng Gao (2016). Summary of metallurgical technology for the treatment of niobium and tantalum ore. Rare metals. 1

[8] Valuable Substances in e-waste, http://ewasteguide.info/valuable_materials_in_e_waste, access 26/3/2017.

[9] Katano S (2014). Recovery of Tantalum Sintered Compact from Used Tantalum Condenser Using Steam Gasification with Sodium Hydroxide. APCBEE Procedia. 10: 182-186

[10] The method of tantalum recovery, http://baike.asianmetal.cn/metal/ta/recycling.shtml, access 26/3/2017.

[11] Gurung M (2013). Recovery of gold and silver from spent mobile phones by means of acidothiourea leaching followed by adsorption using biosorbent prepared from persimmon tannin. Hydrometallurgy. 133: pp 84–93.

[12] DSMZ website. https://www.dsmz.de/home.html, access 20/8/2017.

Comparability of Life Cycle Assessments: Modelling and Analyzing LCA Using Different Databases

Matthias Kalverkamp[1] and Neele Karbe[1]
[1]Cascade Use Research Group, Carl von Ossietzky University of Oldenburg, 26111, Oldenburg, Germany
matthias.kalverkamp@uol.de

Abstract

Life Cycle Assessment is a recognized method to assess the environmental impact of a product, of product alternatives and of deviations from different life cycle designs. Life Cycle Assessment studies usually rely on assumptions regarding certain parts of the modelled life cycle such as material extraction or processing. Life Cycle Inventory (LCI) databases support the modelling by providing the necessary inventory data. Further, software solutions such as Umberto, GaBi thinkstep, and Simapro support the modelling but may further affect the study results. Knowledge on how life cycle inventory databases may influence study results, and hence may steer a decision towards a particular database, is limited. This study aims to better understand deviations and potential inconsistencies between inventory databases causing effects on the life cycle assessment results. The production of an electric motor used in electro-mobility applications such as electric vehicles is modelled using GaBi Thinkstep software and three different life cycle inventory databases, namely Ecoinvent, GaBi professional, and the European Life Cycle Database (ELCD). Starting from the ReCiPe single score results, the analysis moves through corresponding endpoint and midpoint indicators to identify reasons for deviations. Despite some limitations, the results show both similarities and differences. Overall, it may be assumed that results deviate to a degree that can change the resulting assessment. However, the used electricity modules and the modelling of only one product (without alternatives) limit the explanatory power to some regard. Further research, in particular the modelling of product alternatives in the same setup though with different inventory databases, is necessary to better understand the effects of different databases on Life Cycle Assessment results and to assist their addressees in the interpretation of the results.

1 Introduction

Life Cycle Assessment (LCA) is a recognized method to assess the environmental impact of a product, of product alternatives and of deviations from different life cycle designs (e.g. reuse and recycling of products) [1]. LCA results can cause long-lasting impacts, especially when resulting in regulations and laws through policy decisions, although LCA results and meaning depend on a great variety of factors also outside the study, such as market dynamics and local production and processing circumstances [2]. Furthermore, policy decisions based on LCA results can cause significant shifts in financial flows, both of public funds and of private capital moving into different areas and hence into different (technical) solutions. The domain of electro mobility is one prominent example for the application of LCA results. The German government invested EUR 1.2 billion to foster electric vehicles (ELVs) through a subsidy of EUR 4000 for each e-vehicle sold since 2016. At the same time, the government supports the extension of existing e-mobility charging infrastructure with EUR 300 million [3].

However, each individual LCA study depends on a set of parameters influencing the outcome and hence the results. In addition, LCA studies usually rely on assumptions regarding certain parts of the modelled life cycle, such as energy inputs or material demand of certain production processes. The choice for a particular Life Cycle Inventory (LCI) database supporting the LCI analysis of a study has a relevant influence on the LCA results. When compiling a LCA study, one should ideally obtain primary data on material and energy inputs. However, some data are not easily available for the study owner or difficult to obtain, for example, on raw material extraction and processing. Even some data on production processes may be out of reach due to information restrictions of processing companies. Therefore, LCA models rely on LCI databases, such as ecoinvent or GaBi professional; ecoinvent being the most popular LCI database used widely across the globe [4]. Furthermore, LCA studies usually use LCA software solutions such as Umberto, GaBi thinkstep, and Simapro and the choice of LCA solution can further affect the results [5].

The different LCI databases contain data from different sources and years representing particular processes, these "average datasets" and the variety of base years and further circumstances can cause great deviations in LCA results [2]. For e-mobility, especially the electricity mix in the country of use has great impacts during the use phase. Similarly, the electricity mix influences the results of other product life phases requiring energy inputs.

Depending on the (underlying) objectives of a LCA study, the knowledge about differences between the LCI databases may steer a decision towards a particular database. For example, if the focus lies on a certain category of environmental impacts such as water use or fossil depletion, the choice for a LCI database may make an important difference. The overall results of an LCA may also differ significantly between similar models based on different databases. Therefore, this study aims to better understand deviations and potential inconsistencies between LCI databases causing effects on the LCA results. The study asks the questions (1) how much do LCA results deviate because of different LCI databases, and

A. Pehlken et al. (Eds.), *Cascade Use in Technologies 2018*,
https://doi.org/10.1007/978-3-662-57886-5_8

(2) would the choice of a LCI database provide a foreseeable influence on the LCA results, and if so, how could such influence be identified?
To answer the proposed questions, the following section outlines existing literature on LCA software support, on LCI databases, and on their relevance for LCA results. Section 3 explains the methodological approach chosen for this study before section 4 presents the results from the LCAs modelled with different databases. These results are discussed in section 5. This paper concludes with a brief summary of the findings and an outlook on future research opportunities.

2 State of the Art

This section outlines the discussion on inconsistencies between LCA studies that are based on different LCI databases; the resulting potential for manipulation has hardly been addressed in the past but received growing interest in recent years. Some studies investigated these differences and inconsistencies potentially resulting from using different LCA software solutions and especially different LCI databases.
Differences in LCA results depend on different factors. Besides those factors solely relying on the decisions made due to the study design (assumptions, system boundaries, etc.), other factors relate to the software or database used to model and analyse the LCA model. Speck et al. [6] analysed the impact on decision making in packaging depending on the choice of LCA software solution. They compared LCA results of SimaPro and Gabi, two widely used solutions, concluding that neither the software solution itself nor the material amounts but the characterization factors are responsible for deviation in LCA results. These characterization factors depend on the software solution and not on the database; from our experience, characterization factors may be outdated or the implementation of such factors can be inaccurate. Another study on SimaPro and Gabi examined the statistical potential of uncertainties stemming from "statistical value chains" in the biodiesel supply chain. The study concludes that differences in the result derive from errors in the LCI databases; the authors highlight that further research is necessary to verify or falsify the origin of differences [7]. However, a study comparing LCA software solutions (GaBi, SimaPro, openLCA, and Umberto) for a practitioner audience indicates that decisions on software solutions may not consider such observations on the reliability of software solutions and their impact on the LCA results [8].
Although the chosen software solution may influence the LCA results due to issues with the characterization factors, the software solution itself may only have a limited impact on the LCA results, especially when characterization models are implemented correctly. This conclusion, however, does not apply when different databases in the same LCA software are used to model the life cycle. Using only one software solution should at least eliminate deviations due to issues with the implementation of characterization models. Garrain et al. [9] compared the LCI database Ecoinvent and the European Life Cycle Database (ELCD) (amongst others) using the data quality indicators of LCA databases to evaluate selected processes of the database. They conclude that the LCA practitioner needs to carefully assess the quality of ELCD data and that their study results may vary if applied to other than the selected datasets from the LCI database. A further study built upon the latter study to develop a methodological approach for database assessment [10]. A statistical analysis of the LCI databases Ecoinvent and GaBi and the construction-specific databases IBO, CFP and Synergia analysed the deviations between LCA results and the environmental impact in terms of Greenhouse Gas (GHG) emissions of material production. The study identified the allocation rules, the system boundaries, and temporal geographical representativeness of selected modules as major causes for different LCA results [11]. Quandel [12] explicitly addressed the manipulation potential in ecological assessment and investigated how a LCA can be manipulated to favour particular product alternatives. Based on her results, she proposes solutions to overcome such disadvantages regarding the allocation of environmental impacts of co-products in particular.

3 Methodological Approach

This study uses LCA as described by ISO (ISO 14040) though the objective of identifying inconsistencies between LCI databases requires some adaption of the approach. Instead of using one single LCI database to model the lifecycle, this study models the same lifecycle using three different databases. However, this requires also some flexibility in the models themselves since different LCI databases provide processes differing, for example, in their degree of aggregation (e.g. how many previous processes are considered and whether these processes are comparable).

3.1 LCA Model

This study models the life cycle of an e-motor for ELV applications using three different LCI databases, namely ecoinvent, GaBi professional, and the European Life Cycle Database (ELCD). The data on the electric motor (e-motor) for ELVs stems from a study on electro mobility [13]. Using the LCI corresponding to this e-motor, the here presented study models the production of the e-motor using the two databases Ecoinvent, Gabi and ELCD in the LCA modelling solution GaBi thinkstep. The e-motor provides 25 kW of power and weighs 58 kg; iron makes up 58% of the weight of the motor, copper and aluminium make up 20 % of the weight each. The remaining 2 % represent polyethylene [13]. These numbers regarding the material content should be treated with care as some uncertainties remain [13]. Because of the study's objective, the issue of uncertainties in the material composition is not of greatest concern since the relative results between the models are in the focus of the analyses. This study follows a cradle-to-gate-approach modelling the production of the e-motor, considering the production of raw material, the processing of material, transport, and energy and auxiliary resources. Hence, the model does not consider the use phase of the motor nor any end-of-life consideration such as recycling. These reductions of the system boundaries to "cradle-to-gate" derive from the objective of comparing the impact LCI databases have on LCA results. Reducing the complexity helps the comparison of the models. If the objective were to compare product alternatives instead of LCI impacts, the model should ideally cover all life cycle phases of the product.

Necessary assumptions regarding the resource and energy inputs of production and processing activities stem from the approach of Dietz and Helmers (and Hartard) [13, 14]. For example, the study assumes an injection moulding process for the processing of the polyethylene. Furthermore, the production process requires heat which accounts for approx. 4.16 MJ derived from light oil and 144.54 MJ derived from natural gas [13, 15]. In contrast to Dietz and Helmers [14], the here presented model uses the original electricity mixes provided by the LCI databases, although this accounts for relevant differences (see results and discussion). In addition, transport modes and distances are either defined in the LCIs' process modules or depend on the Ecoinvent Report No. 1 for metal transport in Europe providing standardized transport modes and distances for Europe [16]. Finally, the provision of infrastructure necessary for the production of a product causes further environmental impacts. While Ecoinvent provides datasets for such infrastructure, the other LCI databases do not provide separate datasets though supposedly consider impacts of infrastructure in their materials and production processes [17].

The modelling in GaBi using the three LCI databases requires some adjustments to the models. For example, additional modules such as for processing might be necessary in some cases. The consideration of losses during production is another difference in the modelling approaches. For example, GaBi professional usually considers these losses in its process modules while Ecoinvent requires a manual consideration of such losses. If no primary data is available, loss factors can be used to model losses [18].

3.2 LCI Databases

This section briefly outlines the three LCI databases used in this study. At the time of the study design, we were not aware of the discontinuation of the project in 2018. However, we decided to present the results in order to broaden the picture by including a freely available LCI database.

Ecoinvent

The Ecoinvent database is a commercial database from Switzerland, which has been on the market since 2003 and was one of the first comprehensive LCI databases. Today it is one of the leading inventory databases and contains both Swiss and European data [17]. Currently, Ecoinvent contains more than 13.300 datasets covering a wide range of subject areas [19]. The database contains data from the fields of transport, energy, chemicals, agriculture, building materials, metals and their processing, packaging materials, electronics and waste disposal. The database is mainly based on so-called unit processes. In contrast to accumulated process data, these are data records that do not contain several chains of unit processes. For the sake of comparability, they represent the smallest possible process step [20]. A unique feature and, at the same time, a unique selling point of the Ecoinvent database is the use of infrastructure data sets representing, for example, the construction of a factory needed to manufacture a product [17].

GaBi Professional

The Professional Version of the GaBi database contains currently more than 3778 processes. Extension can supplement the basic version. GaBi professional is currently considered one of the market-leading LCI databases [21]. The GaBi database is mainly used in the field of engineering since it provides the datasets relevant for the automotive industry including according suppliers [17]. The data sources of the GaBi datasets are heterogeneous. Preference is given to primary data from industry. However, in some cases, the database provides secondary data from publications, environmental reports from companies or public statistics [22].

European Life Cycle Database

The European Life Cycle Database (ELCD) originated in a project of the Joint Research Center of the European Commission and was published in 2006 [23]. The data sets are limited to the most important production materials, energy sources, transport and waste treatment. The aim of the project was to freely provide a database for ISO 14040-compliant LCA data sets. The data sets came partly from commercial databases or were provided by scientists. For example, data sets from the GaBi database are represented. Data records from the Ecoinvent database are not present in the ELCD. The ELCD is the smallest database compared to Ecoinvent and GaBi. Moreover, the database project was not continued [24].

4 Results

4.1 LCIs Using the Three Different Databases

The e-motor scenario is modelled using three different LCI databases. For each model, dataset modules are used exclusively from the corresponding LCI database. However, some modules might be similar or may even have the same name, especially between GaBi and ELCD. This derives from the fact that ELCD uses some of GaBi's datasets; these datasets are usually older versions compared to the most recent version of GaBi professional. Table 1 summarizes the selected process modules of the three different LCI datasets. Modules marked (*) indicate that additional processing modules might be necessary for proper modelling.

Table 1. LCI Databases and corresponding modules for processes or materials

Material/Process	ELCD	Ecoinvent	GaBi DB
Iron*	Steel rebar (GLO)	Cast iron production (RER)	BF Steel billet / slab / bloom (DE) + Cast iron part (automotive) (DE)
Aluminum*	Aluminium sheet (RER)	Aluminium production, primary, ingot (UN-EUROPE)	Aluminium ingot mix (2010) (EU-27) + Aluminium die-cast part (DE)
Copper*	Copper sheet (EU-25)	Copper production, primary (RER) +	Copper sheet mix (EU-27)
Polyethylen*	Polyethylene low density granulate (PE-LD) (RER)	Polyethylene production, low density, granulate (RER)	Polyethylene low density granulate (LDPE/PE-LD) (DE) + Plastic injection moulding part (unspecific) (DE)
Electricity	Electricity grid mix 1-60kV (EU-27)	Market for electricity, medium voltage (DE)	Electricity grid mix 1kV-60kV (ENTSO)
Heat (light oil)	Heat (EU-27)	Heat production, light fuel oil, at industrial furnace 1MW (Europe without Switzerland)	Thermal energy from light fuel oil (LFO) (EU-27)
Heat (natural gas)	Heat (EU-27)	Heat production, natural gas, at industrial furnace >100kW (Europe without Switzerland)	Thermal energy from natural gas (EU-27)
Water	Process water (EU-27)	Tap water, at user (RER)	Tap water (EU-27)
Transport by truck	Articulated lorry (40t) incl. fuel (RER)	Transport, freight, lorry 16-32 metric ton, EURO3 (RER)	Truck (GLO) + Diesel mix at filling station (EU-27)
Transport by train	Rail transport incl. fuel (EU-27)	Transport, freight train, diesel (Europe without Switzerland)	Rail transport (EU-27)
Vehicle plant	-	Road vehicle factory production (RER)	-

4.2 Ecoinvent-Based LCA Model

Ecoinvent modelling is designed to consolidate processes aiming towards one final process module. Therefore, it is not possible to "pass" a product through a process. This modelling approach reduces the ability to visualise the typically sequential character of a production process (see difference between Figure 1 and Figure 2). Figure 1 shows the Ecoinvent-based model [13, 14] realized in GaBi Thinkstep. For easer orientation and for better comparability with the other scenario models, modules were clustered relating to the corresponding categories of material production, material processing, energy and auxiliary resources, and transport and infrastructure. All process modules connect to the manually created process for aggregation and manufacturing of the e-motor.

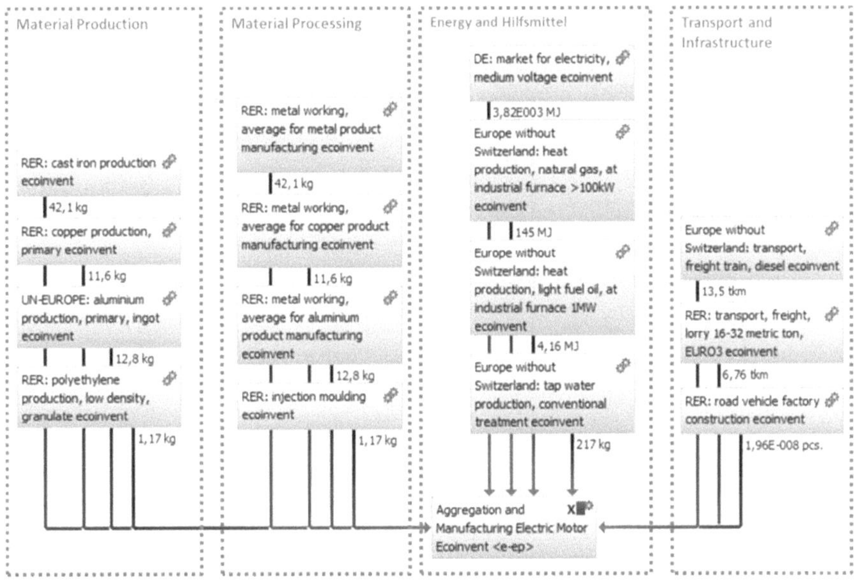

Figure 1. The system modelled in GaBi thinkstep using Ecoinvent

4.3 GaBi-Based LCA Model

The GaBi professional-based model follows a sequential design since this LCI allows for a more process-oriented modelling (see Figure 2). The model in Figure 2 uses almost the same clusters as in Figure 1 (Ecoinvent), which makes these two models easily comparable. When compared to the Ecoinvent-model (Figure 1), the identification of raw materials and corresponding material processing steps is much easier, such as the "EU-27: Aluminum ingo mix (2010) EAA" provides the input material for the "DE: Aluminum die-cast parts ts" process. In Gabi, these processes require additional inputs such as for energy or electricity. For these inputs, corresponding regions can be selected separately (e.g. electricity from different grids; not applicable in this particular case). The modelling approach of GaBi professional results in processes such as "aggregation of materials electric motor GaBi" merging all semi-finished products for rail transport. In this case, a standard distance is assumed. However, if the different semi-finished parts were to be transported over different distances and with different modes of transport, corresponding processes could be easily modelled. The transport cluster also shows that some differences occur within GaBi professional. While the truck transport module provides input connections for "goods" and requires an additional input of diesel fuel, the described aggregation process is still necessary for the rail transport. Both transport processes calculate their energy demand based on the weight of the transported goods. Noteworthy, because of this modelling approach, the transport solely considers those parts for final assembly weighing less than the semi-finished products. Since the transport has a relatively small or almost no influence on the environmental impact of this particular case, this can be neglected here.

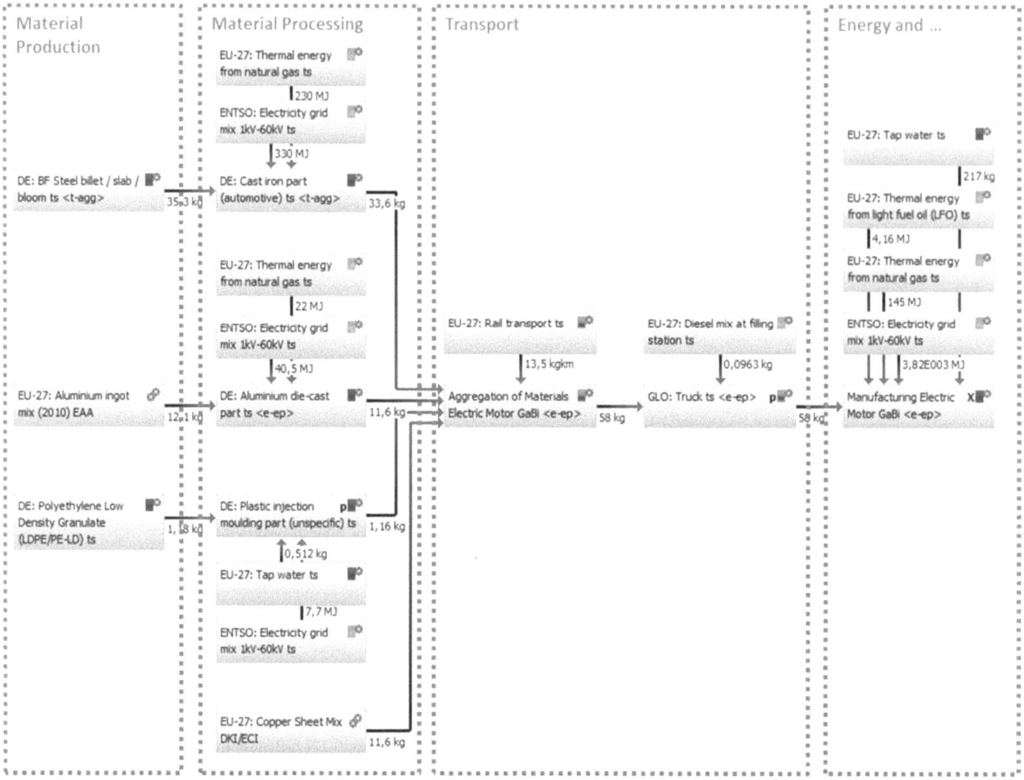

Figure 2. The system modelled in GaBi thinkstep using GaBi professional

Another difference in the modelling approaches is the consideration of losses. Since GaBi professional considers losses in its processing modules, losses are not calculated and hence not added manually to the model. For example, the "plastic injection moulding part (unspecific)" requires 1.18 kg of input material to produce 1.16 kg of output material. One limitation of this GaBi professional-based model is the "copper sheet" process because it represents a type of copper sheet that is usually too thick for applications in electrics. However, the LCI database does not provide a more suitable alternative. This is one issue of LCI databases regardless of their comparability constantly challenging LCA modelers.

4.4 ELCD-Based LCA Model

The ELCD database is the least detailed LCI in this comparison making the modelling more difficult than with GaBi professional or Evoincent. The ELCD also allows for a more process-oriented modelling approach (Figure 3). Nevertheless, the modelling rather appears like a mix of the other two databases as outlined in the following. Once again, a manually added process aggregates all materials ("Aggregation of Material Electric Motor ELCD") considering the truck transport as well. Because of limitations in the software or the ELCD, not more than one transport process with an output flow named "Transport ELCD" can be connected to the same process module. Therefore, the rail transport is connected to the "Manufacturing Electric Motor" process. In the succeeding manufacturing process, loss factors such as in the Ecoinvent-based model were considered. Therefore, the material input at this stage is higher compared to the GaBi professional-based model. Compared to the other two LCI databases, ELCD does not provide as many varieties regarding the regional specificity of processes (such as EU and US, or even country-based) resulting, for example, in the "steel rebar" process being a global ("GLO") process while "Aluminum sheet" is an EU (RER) process. Although the selected material processes seem to consider some production activity as well, losses are considered in addition since these losses refer to, for example, punching and cutting of such material sheets.

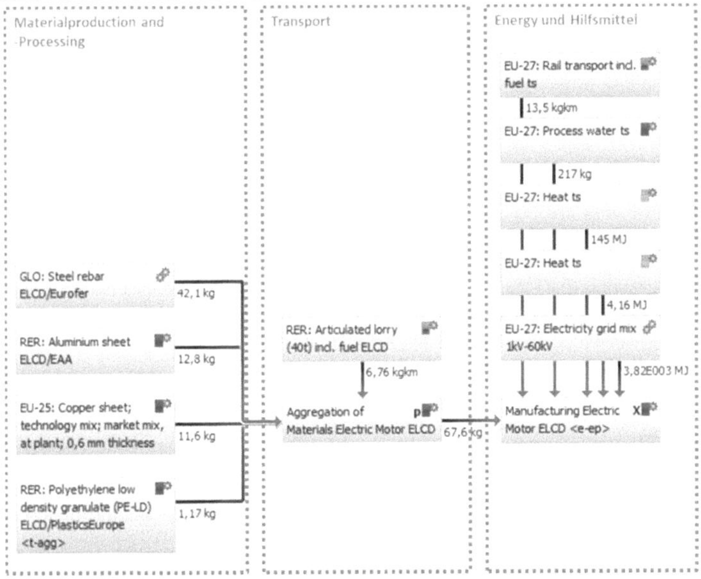

Figure 3. The system modelled in GaBi thinkstep using ELCD

4.5 Impact Assessment with ReCiPe

The impact assessment uses the ReCiPe method. Because of the great variety of midpoint indicators, only a selection will be discussed in detail. The selection is based on their relevance for the use case and the study's objective. The selection covers the ReCiPe midpoint indicators "climate change", "fossil depletion" and "metal depletion". Furthermore, the results cover all endpoint indicators and the single scores. The impact assessment makes use of the clusters that were used to organize the process modules (Material Production, Material Processing, Energy Demand and Auxiliary Resources, and Transport and Infrastructure) as far as possible depending on the corresponding LCI database. Energy required for partly aggregated processing modules (see GaBi model) will be accounted to manufacturing for comparability reasons.

Midpoint: Climate Change

The midpoint indicator Climate Change reflects the negative environmental impact of anthropogenic warming of the Earth's atmosphere [17]. Figure 4 shows the results of the corresponding ReCiPe indicator for the three LCI models. For the Ecoinvent-based model, this indicator accounts for 1114.12 kg CO2-equiv. and corresponding to the outlined process groups, the major contributors are material production (19.27 %), material processing (16.28%), energy demand and auxiliary resources (63.83%). For the ELCD-based model, this indicator accounts for 603.65 kg CO2-equiv. and corresponding to the outlined process groups, the major contributors are material production (16.02%), and energy demand and auxiliary resources (83.98%). For the GaBi professional-based model, this indicator accounts for 717.97 kg CO2-equiv. and corresponding to the outlined process groups, the major contributors are material production (26. 66%), material processing (9.13%), and energy demand and auxiliary resources (64.16%).

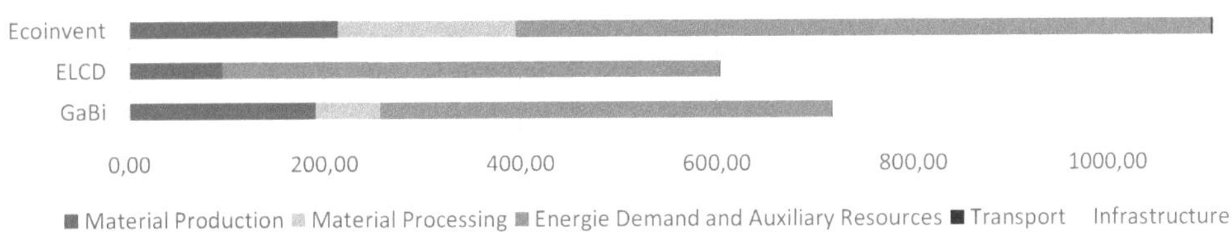

Figure 4. ReCiPe Midpoint Indicator Climate Change (kg CO_2-Equiv.) for the three models

Midpoint: Fossil Depletion

For the Ecoinvent-based model, this indicator accounts for 260.68 kg oil-equiv. and corresponding to the outlined process groups, the major contributors are material production (18.63%), material processing (15.66%), energy demand and auxiliary resources (64.92%). For the ELCD-based model, this indicator accounts for 164.12 kg oil-equiv. and corresponding to the outlined process groups, the major contributors are material production (15.65%), and energy demand and auxiliary resources (84.35%). For the GaBi professional-based model, this indicator accounts for 186.06 kg oil-equiv. and corresponding to the outlined process groups, the major contributors are material production (26.09%), material processing (9.77%), and energy demand and auxiliary resources (64.09%).

Midpoint: Metal Depletion
For the Ecoinvent-based model, this indicator accounts for 799.98 kg Fe-Equiv. and corresponding to the outlined process groups, the major contributors are material production (69.73%), material processing (28.83%), energy demand and auxiliary resources (0.94%). For the ELCD-based model, this indicator accounts for 35.21 kg Fe-Equiv. and corresponding to the outlined process groups, the major contributors are material production (94.19%), and energy demand and auxiliary resources (5.81%). For the GaBi professional-based model, this indicator accounts for 95.52 kg Fe-Equiv. and corresponding to the outlined process groups, the major contributors are material production (96.63%), material processing (0.32%), and energy demand and auxiliary resources (3.06%).

Endpoints and Single Score

Endpoint: Human Health
For the Ecoinvent-based model, the Human Health indicator (in points) accounts for 22.06 points and corresponding to the outlined process groups, the major contributors are material production (19.48%), material processing (17.84%), and energy demand and auxiliary resources (61.34%) (in process order). Transport and infrastructure account for 0.19% and 1.14% respectively. For the ELCD-based model, this indicator accounts for 10.77 points. The major contributors are material production (15.8%), and energy demand and auxiliary resources (84.2%). The remaining process groups do not contribute to this endpoint indicator. For the human health endpoint indicator of the GaBi-model accounting for 13.03 points, the major contributors are material production (26.03%), material processing (9.11%), and energy demand and auxiliary resources (64.81%). The transport (0.06%) and infrastructure (0%) have no or almost no influence.

Resources
For the Ecoinvent-based model, the Resources indicator (in points) accounts for 110.76 points and corresponding to the outlined process groups, the major contributors are material production (46.85%), material processing (22.93%), energy demand and auxiliary resources (29.58%). Transport and infrastructure account for 0.12% and 0.52% respectively. For the ELCD-based model, this indicator accounts for 28.83 points. The major contributors are material production (22.33%), and energy demand and auxiliary resources (77.67%). The remaining process groups do not contribute to this endpoint indicator. For the resources endpoint indicator of the GaBi-model accounting for 36.56 points, the major contributors are material production (38.93%), material processing (8.04%), and energy demand and auxiliary resources (52.98%). The transport (0.05%) and infrastructure (0%) have no or almost no influence.

Ecosystem Quality
For the Ecoinvent-based model, the Ecosystem Quality indicator (in points) accounts for 112.64 points and corresponding to the outlined process groups, the major contributors are material production (19.06%), material processing (16.43%), energy demand and auxiliary resources (64.11%). Transport and infrastructure account for 0.07% and 0.32% respectively. For the ELCD-based model, this indicator accounts for 15.21 points. The major contributors are material production (16.6%), and energy demand and auxiliary resources (83.4%). The remaining process groups do not contribute to this endpoint indicator. For the resources endpoint indicator of the GaBi-model accounting for 19.05 points, the major contributors are material production (30.45%), material processing (10.47%), and energy demand and auxiliary resources (59.04%). The transport (0.04%) and infrastructure (0%) have no or almost no influence.

Single Score
Figure 5 shows the Singe Score results (in points) corresponding to the three LCI models. The Singe Score consist of the three endpoint indicators "Human Health", "Resources" and "Ecosystem Quality". For Ecoinvent, the Single Score amounts to 235.46 points. With regard to the self-selected material groups, the result amounts to 30.99% for the material production modules, to 19.35% for the material processing modules, to 49.08% for the energy and auxiliary resources modules, to 0.1% for the transport modules and to 0.48% for the infrastructure modules (Ecoinvent only). The ELCD-based model amounts to 54.82 points in the single score, the lowest amongst the three models. With regard to the self-selected material groups, the result amount to 19.46% for the material production modules, and to 80.54% for the energy and auxiliary resources modules. The remaining groups do not contribute to the single score. The GaBi-based model mounts to a Single Score of 68.63 points. With regard to the self-selected material groups, the result amounts to 34.12% for the material production modules, to 8.92% for the material processing modules, to 56.91% for the energy and auxiliary resources modules, and to 0.05% for the transport modules. The infrastructure module is not Ecoinven-only hence does not contribute to the "GaBi Single Score".

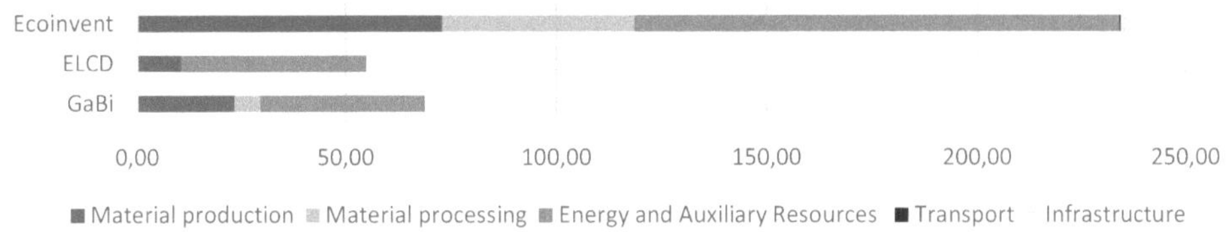

Figure 5. ReCiPe Single Score (points) for the three models

Sensitivity Analysis

For the sensitivity analysis, electricity was selected because of the overall relevance of the electricity and auxiliary resources module groups. Each model was therefore re-calculated using renewable energy modules (RE) instead of the regular energy modules (Electriciy Mix). For each model, energy modules were selected corresponding to the LCI database. For the Ecoinvent-based model, the „market for electricity, medium voltage, label certified (CH)" substitutes the „market for electricity, medium voltage (DE)" module (unfortunately, there was no DE-module for renewable energy available). For the ELCD-based model, the „Electricity from wind power (RER)" module substitutes the "Electricity grid mix 1kV-60kV (EU-27)" module. For the GaBi-based model, the „Electricity from wind power (EU-27)" process substituted the „Electricity grid mix 1kV-60kV (ENTSO)" process. The single score is used to compare the results (Figure 6). On average, the single score results are 61% below the regular results (substituted energy modules). With 77% below the reference value, the difference is highest in the case of ELCD. However, the ELCD model uses the least individual modules hence changing one single module may have a greater relative impact of the single score than in the other cases. Regardless of this limitation, the results show that the energy type used for production and processing has a significant impact on the LCA results in each of the three cases.

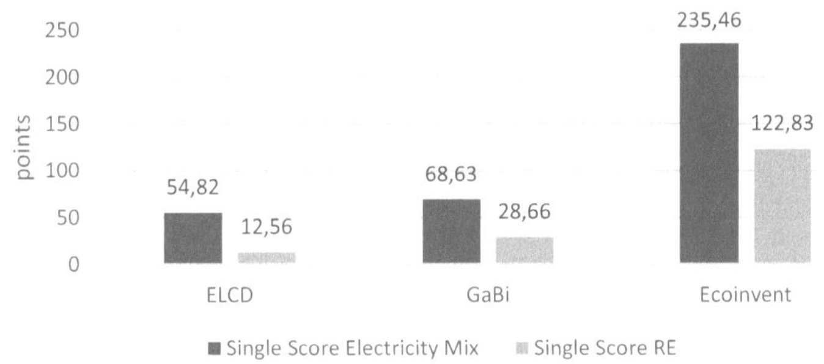

Figure 6. Single Score Results for the Sensitivity Analysis

4.6 Comparison of Overall Results

Table 2 shows the Life Cycle Impact Assessment Results based on ReCiPe for selected Midpoint indicators as well as for the Endpoint indicators and the Single Scores (both regular models and from the sensitivity analysis). It is noticeable that all impact indicators are higher in the case of the Ecoinvent database except for the impact indicator "Ionizing Radiation", which is lower compared to the results of the other two databases. In the case of the GaBi database as well as the ELCD, the electricity module "Electricity Grid Mix 1kV-60kV (ENTSO)" contributes just below 84% to the total amount of the ionizing radiation effect indicator. When modeling using the ELCD, 98% of the overall result stem from the environmental impact of the Electricity Grid Mix 1kV-60kV (EU-27) electricity module. The result of the Ecoinvent modeling is also dominated by the electricity module "market for electricity, medium voltage (DE)", but only by approximately 71%.

Table 2. Life Cycle Impact Assessment (ReCiPe) Results for selected Midpoints, Endpoints, the Single Score and the Single Score of the Sensitivity Analysis

	Ecoinvent	ELCD	GaBi
Human Toxicity (kg 1.4-DB Equiv.)	2012.48	18.71	74.94
Climate Change (kg CO2-Equiv.)	1114.12	603.65	717.97
Photochemical Oxidation (kg NMVOC)	2.97	1.51	1.69
Fossil Depletion (in kg oil-Eqiv.)	280.68	164.12	186.06
Metal Depletion (in kg Fe-Equiv.)	799.98	35.21	95.52
Human Health (points)	22.06	10.77	13.03
Resources (points)	100.76	28.83	36.56
Ecosystem Quality (points)	112.64	15.21	19.05
Single Score (points)	235.46	54.82	68.63
Single Score (Sensitivity Analysis) (points)	122.83	12.56	28.66

Similar to Takano et al [11], the standard deviation is used to identify which ReCiPe indicator has the largest differences between the different databases. With a relative standard deviation of 162%, the values of the human toxicity midpoint indicator are most different between the three LCAs. Terrestrial ecotoxicity follows with 142% and the consumption of metals with 137%. The ecosystem quality endpoint indicator shows a standard deviation of 113%, the single score of 84% and the endpoint indicator resources of 71%. The human health and photochemical oxidation indicators (39%), the climate change indicator (33%) and the fossil depletion indicator (29%) show the lowest deviations.

4.7 Deviations of Selected Processes between the Three Models

The single score results summarizing the endpoint results serve as basis to identify and analyse selected process modules that may have a significant impact on the deviation of the LCA results. The single score results of individual modules (not of the entire LCA) show similarities and differences as well. Comparing material modules from the three databases, the polyethylene modules (without processing modules) result in very similar single scores (Ecoinvent and GaBi 0.4, ELCD 0.41 [points]). Similarly, the Ecoinvent and the GaBi professional database modules for iron result in single scores of 11.3 (GaBi) and 11.73 (Ecoinvent) (points). Greater deviations occur for aluminium and copper. For aluminium, Ecoinvent scores 1.5 times higher than GaBi professional does while ELCD scores less than 0.5 times of GaBi professional. Especially copper shows very great deviations with a single score of 45.12 points for Ecoinvent, this score is 29 times higher than the corresponding GaBi professional score and still 20 times higher than the ELCD score.

On average, the single scores of the ELCD material modules are only half the corresponding GaBi single scores and less than 15% of the corresponding Ecoinvent scores. The single scores of the GaBi professional material modules are on average 1/3 of the ecoinvent scores. This observation of single scores does not include the processing modules (such as injection moulding). This is important because the Gabi professional processing modules account for much less additional points regarding the single scores than its counterparts in the Ecoinvent-based model. The aluminium processing in the Ecoinvent model is an extreme case but illustrates the impact: The Ecoinvent module causes an almost 600 times higher score than the corresponding GaBi module (ELCD does not provide separate processing modules).

When looking at the energy and auxiliary resources, the natural gas modules of all databases provide similar results. However, the electricity modules show differing results. While the electricity modules of GaBi professional and ELCD cause similar single scores, the score of the corresponding ecoinvent module is significantly higher and at the same time the highest single score value of all modules. The Ecoinvent transport modules score more than 10 times then the other two databases whose single scores are 0.1 or lower.

In summary, in all three models the electricity modules cause the highest impact to the single score. In case of GaBi professional and ELCD, the material module for iron causes the next highest contribution to the single score, followed by aluminium. In case of Evoinvent, the material module for copper is second highest regarding its contribution to the single score, followed by iron. The latter would be different if processing modules were considered as well. In that case, aluminium would be second highest contributor. These processing modules are an important contributor to the LCA results in case of Ecoinvent. As a result, the high environmental impact of the processing modules in case of Ecovinent contributes significantly to the different database-specific singe scores.

5 Discussion

Based on the LCI database choice, significant differences in LCA results may occur. Reasons for deviations are different modelling approaches underlying the design of the respective LCI database. In this study, we focused on the LCI databases used with one particular software solution, namely GaBi. The GaBi software is designed for a more process-oriented modelling approach as is the LCI database GaBi professional. However, different LCI databases can be used with the GaBi software, such as ecoinvent and ELCD. Due to the discontinuation of ELCD, corresponding results are not discussed in detail. In addition to the LCI database, the selection of a LCA software solution (such as GaBi) may have an influence on the study results. For example, GaBi thinkstep advises the user that datasets from Ecoinvent used in GaBi may cause unrealistically high results for water consumption [25]; the here presented study did not show any such issue. However, errors and deviations due to compatibility issues between software and databases may not be eliminated and hence may provide some margin for manipulation.

One prominent example for differences in LCA results due to different LCI database relating to e-mobility can be found in the energy mix provided by LCI databases. As the latest available energy mix for EU/Germany, the LCI Ecoinvent V 3.1 energy mix is based on the year 2008 while the LCI GaBi professional energy mix is based on the year 2012. Since the energy mix has changed over the years 2008-2012, the results of a LCA using this energy mix, for example as energy input during the use phase of an ELV, depend significantly on the available energy mix. In addition, the results depend on the chosen model for the data interpretation such as ReCiPe or Ecoindicator 99. Due to this significant variation between LCI databases regarding the electricity mix, the sensitivity analysis performed had only limited explanatory power. Nevertheless, it shows that LCA requires specialized knowledge also on the LCI database. Users of LCI databases may not always be aware of particular details causing great impacts on the LCA results. This is both a limitation of the here presented study as it is a result, namely the disadvantages when almost entirely relying on data from LCI database for the life cycle inventory assessment (LCIA). The study showed that not only the electricity mix, although having a significant impact on the results, but also material production and processing causes important deviations. This may in fact be more worrying than the electricity mix. While the electricity mix can be individually designed using electricity production modules for different energy sources, these production modules are much harder to replace with primary data by LCA practitioners.

Another reason for deviations in LCA results between software solutions (GaBi thinkstep, Umberto, Simapro, etc.) may be difficulties in adapting the available LCI databases to these software solutions. This aspect has not been addressed by the analysis of the results. Since the modelling approaches of LCA software solutions differ and since the adaption of impact assessment models to software solutions provides some room for human error, this aspect of LCA should be of concern for every LCA practitioner and the research community.

The results of the different models showed significant deviations and accordingly address the first research question asking how much LCA results deviate because of different LCI databases. Although this study only discusses one case using three different LCI databases, and hence results are limited to this case, other studies have come to similar results. Therefore, and since no other findings are known, it may be assumed that results deviate to a degree that can change the resulting assessment. However, it has not been assessed in this case whether the assessment of product alternatives (e-motor vs combustion engine) would result in contradicting product-alternative suggestions when using different databases. Such assessment would provide additional insight on the manipulation potential through LCI database choices. The latter leads to the second research question that asked if the choice for a LCI database would provide foreseeable influence on the LCA results, and if so, how to identify such influence? The results of this study do not provide sufficient evidence for foreseeable influence on LCA results depending on the applied LCI database. Nevertheless, the result for this case indicate that the relative contributions of particular product life phases (production, processing) may differ and that the different databases assess the impact of certain modules differently. Specific knowledge on these differences may provide some margin for manipulation, for example when setting system boundaries or when selecting particular modules such as aggregated modules for products instead of separate production and processing models. Such choice may even be argued satisfyingly hence not raising immediate concern when analyzing the LCA results. It should be noted that the intention of this study is not to motivate for such manipulation but rather to raise awareness for such issues beyond the LCA professionals who most likely are aware of such challenges.

6 Conclusion and Outlook

The results showed that the decision for a LCI database can have a significant impact on the LCA result hence on potential decisions made based on such results. Different LCA stakeholders may have different objectives and intentions when conducting LCA studies or when using study results. Therefore, addressees of LCA results have to take great care when provided with LCA study results, for example as decision support. This is, however, a challenge especially for stakeholders not being experts on conducting LCA studies.
The electricity modules proved to cause significant impact on the overall results and at the same time showed significant deviations between the three databases. In all three models, the chosen electricity modules provide 1 kWh of electricity in a medium voltage grid to the end consumer. All modules consider losses of the electricity transmission and voltage transformation. An analysis of these electricity modules would be promising to generate a better understanding for deviations in the results. For example, the (output) mass flows of each individual module could be compared. Although even such analysis has its limitation since databases may have their unique flows, it would contribute to the knowledge on LCIA and the challenges when using LCI databases in general.
This research paved the way towards a forthcoming study on LCI database comparison using a life cycle model including the use phase and a more detailed analysis of the LCA results.

7 Zusammenfassung

Die Ökobilanzierung ist eine anerkannte Methode zur Bewertung von Umweltauswirkungen eines Produkts, von Produktalternativen und von Abweichungen bei unterschiedlichen Lebenszyklusdesigns. Life-Cycle-Assessment-Studien basieren in der Regel auf Annahmen in Bezug auf bestimmte Teile des modellierten Lebenszyklus wie Materialgewinnung oder -verarbeitung. Life Cycle Inventory (LCI) -Datenbanken unterstützen die Modellierung, indem sie die erforderlichen Inventardaten bereitstellen. Darüber hinaus unterstützen Softwarelösungen wie Umberto, GaBi thinkstep und Simapro die Modellierung, können aber die Studienergebnisse weiter beeinflussen. Das Wissen darüber, wie Daten aus Life Cycle Inventory-Datenbanken die Studienergebnisse beeinflussen können, könnten somit auch eine Entscheidung für eine bestimmte Datenbank steuern. Diese Studie zielt darauf ab, Abweichungen und mögliche Inkonsistenzen zwischen Inventardatenbanken, die Auswirkungen auf die Ökobilanz-Ergebnisse haben, bessere zu verstehen. Die Produktion eines Elektromotors, der in Elektromobilitätsanwendungen wie Elektrofahrzeugen verwendet wird, wird mit der Software GaBi Thinkstep und drei verschiedenen Bestandsdatenbanken (Ecoinvent, GaBi professional und der European Life Cycle Database (ELCD)) modelliert. Ausgehend von den ReCiPe-Einzelergebnis-Ergebnissen bewegt sich die Analyse von den entsprechenden Endpoint- zu den Midpoint-Indikatoren, um Gründe für Abweichungen zu identifizieren. Trotz einiger Einschränkungen zeigen die Ergebnisse sowohl Ähnlichkeiten als auch Unterschiede. Insgesamt kann davon ausgegangen werden, dass die Ergebnisse zu einem Grad abweichen, der die Beurteilung beeinflussen kann. Die verwendeten Strommodule und die Modellerstellung nur eines Produkts (ohne Alternativen) schränken jedoch die Aussagekraft einigermaßen ein. Weitere Untersuchungen, insbesondere die Modellierung von Produktalternativen im selben Setup, jedoch mit unterschiedlichen Inventardatenbanken, sind notwendig, um die Auswirkungen unterschiedlicher Datenbanken besser zu verstehen und um Empfänger von Ökobilanzen bei der Interpretation zu unterstützen.

Acknowledgments. Matthias Kalverkamp was financially supported by the German Federal Ministry of Education and Research (BMBF) in the Globaler Wandel Research Scheme (Grant No. 01LN1310A). Furthermore, the authors are grateful to Johannes Dietz and Eckard Helmers for providing insights into the modelling of their study; the latter being the basis for a joint study on LCI database comparison (forthcoming).

8 References

[1] Guinée JB, Heijungs R, Huppes G et al. (2011) Life cycle assessment: Past, present, and future. Environ Sci Technol 45(1): 90–96. doi: 10.1021/es101316v

[2] Krozer J, Vis JC (1998) How to get LCA in the right direction? Journal of Cleaner Production 6(1): 53–61. doi: 10.1016/S0959-6526(97)00051-6

[3] BMWi (2016) BMWi - Rahmenbedingungen und Anreize: Bundesministerium für Wirtschaft. Accessed 02 Sep 2016

[4] Rodríguez C, Ciroth A (2016) Adaption LCA software to LCI databases and vice versa. Ökobilanzwerkstatt, Pforzheim

[5] Speck R, Selke S, Auras R et al. (2016) Life Cycle Assessment Software: Selection Can Impact Results. Journal of Industrial Ecology 20(1): 18–28. doi: 10.1111/jiec.12245

[6] Speck R, Selke S, Auras R et al. (2015) Choice of Life Cycle Assessment Software Can Impact Packaging System Decisions. Packag. Technol. Sci. 28(7): 579–588. doi: 10.1002/pts.2123

[7] Herrmann IT, Hauschild MZ (2012) Improving life cycle assessment methodology for the application of decision support: Focusing on the statistical value chain. Dissertation. Technical University of Denmark (DTU), Kgs. Lyngby

[8] Lüdemann L, Feig K (2014) Vergleich von Softwarelösungen für die Ökobilanzierung: Eine softwareergonomische Analyse. Logistics-journal.de. https://www.logistics-journal.de/not-reviewed/2014/09/3991. Accessed 09 Sep 2016

[9] Garraín D, Fazio S, La Rúa C de et al. (2015) Background qualitative analysis of the European reference life cycle database (ELCD) energy datasets - part II: Electricity datasets. Springerplus 4: 30. doi: 10.1186/s40064-015-0812-2
[10] Fazio S, Garraín D, Mathieux F et al. (2015) Method applied to the background analysis of energy data to be considered for the European Reference Life Cycle Database (ELCD). Springerplus 4: 150. doi: 10.1186/s40064-015-0914-x
[11] Takano A, Winter S, Hughes M et al. (2014) Comparison of life cycle assessment databases: A case study on building assessment. Building and Environment 79: 20–30. doi: 10.1016/j.buildenv.2014.04.025
[12] Quandel AM (2015) Ökologische Bewertung. Manipulationsspielräume und ihre Eingrenzung. Dissertation, RWTH Aachen
[13] Dietz J, Helmers E (2015) Ökobilanzierung von Elektrofahrzeugen: Abschlussbericht im Modul 8a, Teil A
[14] Helmers E, Dietz J, Hartard S (2017) Electric car life cycle assessment based on real-world mileage and the electric conversion scenario. Int J Life Cycle Assess 22(1): 15–30. doi: 10.1007/s11367-015-0934-3
[15] Notter DA, Gauch M, Widmer R et al. (2010) Contribution of Li-ion batteries to the environmental impact of electric vehicles. Supporting Information. Environ Sci Technol 44(17): 6550–6556. doi: 10.1021/es903729a
[16] Fischknecht R, Jungbluth N (2007) Overview and Methodology: Ecoinvent Report No. 1, Dübendorf
[17] Klöpffer W, Grahl B (2009) Ökobilanz (LCA) - Ein Leitfaden für Ausbildung und Beruf. Wiley-VCH, Weinheim
[18] Habermacher F (2011) Modelling Material Inventories and Environmental Impacts of Electric Passenger Cars: Com pariosn of LCA results between electric and conventional vehicle scenarios. Master Thesis,, ETH Zurich
[19] Ecoinvent Why ecoinvent. http://www.ecoinvent.org/database/buy-a-licence/why-ecoinvent/why-ecoinvent.html. Accessed 12 Jun 2018
[20] Hischier R (2011) ecoinvent - eine konsistente, transparente und qualitätsgesicherte Hintergrunddatenbank für Ökobilanzen & Co. Chemie Ingenieur Technik 83(10): 1590–1596. doi: 10.1002/cite.201100137
[21] Thinkstep (2018) GaBi LCA Datenbanken. http://www.gabi-software.com/deutsch/databases/gabi-databases/. Accessed 12 Jun 2018
[22] Baitz M, Makishi Colodel C, Kupfer T et al. (2014) GaBi Database & Modelling Principles 2014. http://www.gabi-software.com/fileadmin/gabi/Modelling_Principles/GaBi_Modelling_Principles_2014.pdf. Accessed 12 Jun 2018
[23] European Commission (2013) Background analysis of the quality of the energy data to be considered for the European Reference Life Cycle Database (ELCD). http://eplca.jrc.ec.europa.eu/uploads/Final-Report-CIEMAT.pdf. Accessed 12 Jun 2018
[24] European Commission (2018) ELCD Discontinuation. http://eplca.jrc.ec.europa.eu/ELCD3/. Accessed 12 Jun 2018
[25] Thinkstep Lifecycle Assessment LCA Software: GaBi Software. http://www.gabi-software.com. Accessed 19 Aug 2016

Reuse, Recycling and Recovery of End-of-Life New Energy Vehicles in China

Weiqun Han[1, 2], Yuan Shi[2], Alexandra Pehlken[3], Goufang Zhang[2], Pang-Chieh Sui[2], Jinsheng Xiao[2]
[1]College of Science & Arts, Jianghan University, Hubei, 430056, China
[2]Hubei Key Laboratory of Advanced Technology for Automotive Components and Hubei Collaborative Innovation Center for Automotive Components Technology, Wuhan University of Technology, Hubei, 430070, China
[3]Cascade Use Research Group, Carl von Ossietzky University, Oldenburg, 26129, Germany
alexandra.pehlken@uni-oldenburg.de, jinsheng.xiao@whut.edu.cn

Abstract

China has the largest quantity and ownership of new energy vehicles (NEVs) in the world. As time progresses, a certain percentage of NEVs will enter the scrap stage every year in China. By examining the status quo and the problems of recycling conventional end-of-life vehicles in China, this study investigated the potential market for end-of-life NEVs and the existing problems in the recycling and reuse systems. It was found that the recycling and reuse systems' needs should be considered ahead of time. Further, various modes and specific methods for the recycling and reuse of end-of-life NEVs and their parts were analyzed. Based on this investigation, several strategies for the recycling and reuse of NEVs are proposed. As a future recycling model for end-of-life NEVs, China should establish an independent industry model for the comprehensive utilization of scrap power batteries from NEVs. This study provides a reference for establishing a sound recycling system for end-of-life NEVs and setting up a recycling and reuse platform for scrapped NEVs in China.

1 Introduction

After years of demonstration and promotion, by the end of 2017, the cumulative number of new energy vehicles (NEVs) in China had exceeded 1.7 million, making China the country with the largest number of NEVs in the world. This means that as time passes, a certain number of NEVs and power batteries will enter the end-of-life stage each year in China. Because of this fact, there is an urgent need to examine issues related to the recycling and reuse of NEVs. China's large-scale promotion of NEVs produces a conflicting scenario. Zero emissions can be achieved, which can greatly help to reduce smog and protect the environment; however, if NEVs, and especially power batteries, are improperly processed, this can create new and more serious forms of pollution. Therefore, China's automotive industry must now consider how to properly dispose of end-of-life NEVs, rationally reuse and recycle key components and materials, reduce pollution caused by NEV waste, and promote sustainable development.

Chen et al. [1-5] analyzed the traditional recycling system of China's end-of-life vehicles and highlighted some key issues in the transformation and upgrading of enterprises that oversee end-of-life vehicle recycling and dismantling. After reviewing the relevant policies and standards for China's end-of-life vehicle recycling and parts remanufacturing, they proposed measures to support the development of China's end-of-life auto industry, which could play an important role in China's automobile scrap recycling industry. Shen et al. [6-8] examined the laws and regulations concerning automotive product recycling in China under the extended producer responsibility system. They suggested that the key to building a complete recycling system for automotive products lies in implementing the extended producer responsibility system and improving the recycling and reuse levels of automotive companies. Drawing lessons from the recycling of scrapped automobiles and parts abroad, Li et al. [9-11] examined the recovery of scrapped automobiles and parts in Germany and Japan. They suggested that China should introduce a law for recycling of end-of-life vehicles and their parts as soon as possible to improve the regulatory system and strengthen law enforcement.

The development of China's NEV industry continues to expand. Annual sales have ranked first in the world for three consecutive years, and annual sales of new energy vehicles have accounted for about half of worldwide sales. Regarding the future industrial development of China's NEVs, Wan et al. [12-13] predicted that the sales volume of NEVs would reach 1 million in 2018. Meanwhile, Miao et al. [14] argued that there are some problems in China's new energy automotive industry, such as the lack in the after-sales service support system, which has had certain negative effects on cultivating the consumer market. Han et al. [15] reviewed the demonstration and promotion of fuel cell electric vehicles in China in recent years and proposed several strategies to promote their marketization. Ding [16] and Yu et al. [17] analyzed the development of the recycling and dismantling industry for China's end-of-life NEVs. They noted that the recycling of scrapped NEVs is different from that of traditional cars and therefore cannot continue under the policies and standards for traditional cars. This issue will undoubtedly create a hurdle in the recycling of NEVs.

Many studies have investigated the recovery and reuse of power batteries. A few researchers [18-23] reported an economic analysis model for the input and output of the power battery recycling process and conducted a quantitative analysis. Regarding the cascade use of power batteries, many researchers have noted that it is a desirable concept but difficult to implement. The main reason for such difficulty is that there are too many types of power batteries in China, and the output is scattered. Regarding the recycling system and recovery model for NEV power batteries, Li et al. noted

A. Pehlken et al. (Eds.), *Cascade Use in Technologies 2018*,
https://doi.org/10.1007/978-3-662-57886-5_9

that due to the short time for the marketization of NEVs in China, automotive power batteries have not been scrapped in batch. As such, a domestic power battery recycling system has yet to be commercially formed [24]. Lu et al. [25] suggested that in the future, China might consider establishing a waste power battery recycling network centered on power battery manufacturing companies and new energy vehicle production enterprises. Hou [26] and Peng [27] proposed three models for future power battery recycling: a power battery manufacturer recycling mode, an industry coalition recycling mode, and a third-party recycling mode. Chen [28] simplified the quantitative analysis of the need to recover precious metal materials for proton exchange membrane fuel cells.

Existing research on the recovery of end-of-life automobiles and parts has mainly focused on traditional automobiles. Studies of NEVs have mainly focused on R&D, technology, and demonstration and promotion. While a few studies have investigated the recovery and reuse of NEVs and parts, they mainly focused on one aspect (e.g., recycling modes) or one part (e.g., batteries). Currently, there is a need for research on the recovery and reuse of NEVs and components that is more systematic and conceptual. This article introduces the status quo and problems of China's traditional end-of-life vehicle recycling. Further, it predicts the potential recycling market for China's end-of-life NEVs, followed by a comprehensive and systematic description of recycling modes and methods for China's NEVs and their parts. Finally, several market strategies are proposed to promote the recycling of China's NEVs and their parts and components.

2 Status and Problems of Recycling End-of-Life Conventional Vehicles

2.1 Current Status of Recycling End-of-Life Conventional Vehicles

From 2009 to 2017, China ranked first in the world in vehicle production and sales. In 2017, the annual sales of automobiles in China exceeded 28.88 million, and vehicle ownership reached 217 million. Figure 1 shows the sales volume, growth rate, and ownership of conventional vehicles in China from 2001 to 2017 [29-30]. With this rapid increase in vehicle ownership, the amount of end-of-life vehicles will also greatly increase. Many cities in China cannot dismantle end-of-life vehicles quickly enough, and they pile up in "vehicle graveyards." The annual scrap rate of the international mature automobile market is 5–8% of owned vehicles [2]. From 2018 onward, the annual number of end-of-life vehicles in China is projected to reach 10.85–17.36 million. Vehicle ownership in China was 105 million in 2011, and more than 100 million vehicles have been added in the past six years. The average scrapping time for automobiles is 8–15 years, which means that China's actual scrap amount is currently about six million vehicles. China will face an excess of end-of-life vehicles in five years; such a large amount of scrap is a huge "resource of urban minerals" for China's auto recycling industry.

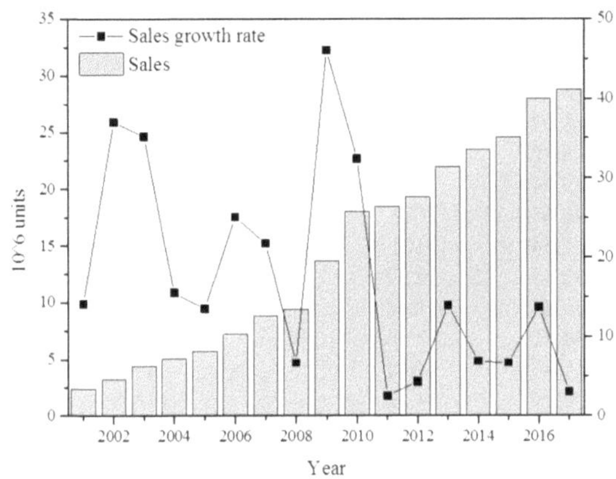

Figure 1. China's automobile sales and annual growth rates during 2001–2017 [29-30]

China's end-of-life vehicle recycling industry started late and has not attracted sufficient attention from the domestic auto industry. In 2001, the Management Rules of Recycling End-of-Life Vehicles were implemented. In 2006, the Technology Policy for Auto Products Recycling was promulgated. Subsequently, China formulated and improved a series of policies and regulations concerning the management of end-of-life vehicle recycling. However, policy implementation has been slow, and the development of the end-of-life automobile dismantling industry lags behind the development of China's automobile industry. China's conventional end-of-life vehicle recycling and dismantling industry has always been a labor-intensive industry with large differences in overall quality. China's enterprises engaged in the large-scale qualified recycling and dismantling of end-of-life vehicles use modern dismantling methods and devices. However, smaller recycling enterprises and illegal recycling enterprises can be relatively backward. Some do not even have special dismantling equipment and basically dismantle vehicles manually, with low efficiency and low social recognition.

Meanwhile, automobile manufacturers rarely consider the recycling performance of their products during the design phase, and the parts recycling rate is low.

The revised version of China's Management Rules of Recycling End-of-Life Vehicles will be introduced soon. It will focus on three reforms: (1) to remove the limit on the total number of end-of-life vehicle recycling and dismantling enterprises, certificates will be issued to these enterprises first, followed by licenses; (2) dismantled parts will be allowed to reenter the market to achieve green vehicle marketing; and (3) giving the "five major assemblies" (engine, transmission, front axle, rear axle, and frame) to qualified remanufacturers for remanufacturing will be encouraged.

2.2 Existing Problems of Recycling Conventional End-of-Life Vehicles

At present, the main problems existing in the recycling of end-of-life vehicles in China are as follows:

The legal scrap rate of end-of-life vehicles is low.

China's 2016 data are taken as an example. According to data released by the Traffic Management Bureau of the Ministry of Public Security of the People's Republic of China, in 2016 vehicle ownership in China reached 194 million, and a total of 5.5 million vehicles had been written off. However, only 1.592 million vehicles were collected by qualified end-of-life vehicle recycling companies. The legal scrap rate was less than 30%. A total of 3.9 million written-off vehicles went to the black market and to second-hand markets; these arrive at remote mountainous regions or rural areas, potentially creating many traffic hazards. Currently, the actual recycling amount recovered by China's qualified end-of-life vehicle recycling enterprises is only 0.5–1.0% of owned vehicles, which is far below the level of 5.0–8.0% in developed countries [2]. Table 1 shows the theoretical estimation of the recovery amount of end-of-life vehicles in China and the actual recovered amount of qualified end-of-life vehicles.

The dismantling level and recovery rate are low.

Most recycling enterprises in China are small, plant sites are crude, and dismantling methods are primitive. Manual processes are often used in these companies, with low dismantling efficiency and great risks to workers' safety. At present, aside from steel and nonferrous metals, most other items cannot be effectively recycled, and the added value of retained parts is low. Parts and components only account for 5–10% of automobile recycling companies' sales, and about 90% are sold as raw materials to steel mills.

Table 1. Prediction of China's end-of-life conventional vehicles and new energy vehicles

Year	Conventional vehicles				New energy vehicles		
	Population (10^6 unit)	Increase (10^6 unit)	End-of-life (10^6 unit)	Recycled (10^6 unit)	Population (10^4 unit)	Increase (10^4 unit)	End-of-life (10^4 unit)
2015	172	24.59	6.16	1.7	49.70	33.11	-
2016	194	28.02	6.88	1.59	100.00	50.70	1.50
2017	217	28.88	7.76	1.47	176.2	77.7	3.0
2018	238	29.74	8.68		279.7	108.8	5.28
2019	255	30.63	9.52		423.6	152.3	8.39
2020	273	31.54	10.2		624.1	213.2	12.7
2021	292	32.48	10.92		848.7	255.8	31.2
2022	314	33.45	11.68		1062.1	307	42.4
2023	336	34.45	12.56		1377.4	368.4	53.1
2024	358	35.48	13.44		1750.5	442	68.9
2025	381	36.54	14.32		2193	530	87.5

Formulae for conventional vehicles:
(1) Increase rate = 3% per year during 2018–2025
(2) Population = (previous population) + increase – (end-of-life)
(3) End-of-life = (previous population) * 4% during 2018–2025
Source: [31]

Formulae for new energy vehicles:
(1) Increase rate = 40% during 2018–2020; 20% during 2021–2025
(2) Population = (previous population) + increase – end-of-life
(3) End-of-life = (previous population) * 3% during 2016–2020; 5% during 2021–2025
Source: [32]

The illegal dismantling and sale of end-of-life vehicles has continued despite prohibitions.

China's Management Rules of Recycling End-of-Life Vehicles clearly stipulates that all companies and individuals are prohibited from assembling vehicles using scrapped cars, and such assembled vehicles and scrapped cars are prohibited from traveling on roads. Due to a lack of effective oversight, small workshops that illegally dismantle automobiles still operate in some localities. Harmful substances such as lead, mercury, cadmium, and hexavalent chromium are freely discharged without treatment, resulting in damage to the environment and risks to human health.

3 Markets of Recycling End-of-Life New Energy Vehicles

3.1 Current Market of Recycling End-of-Life NEVs

The NEV is one of China's seven emerging strategic industries. Developing NEVs is the only way China can transform from a large automobile country into a powerful automobile country. The demonstration and promotion of NEVs in China began 20 years ago with the Shantou Electric Vehicle pilot operation in 1998. Early-stage large-scale promotion projects included the 2008 Beijing Olympic Games, the 2010 Shanghai World Expo, the Guangzhou Asian Games, and the 2011 Shenzhen Universiade, and the "Thousand New Energy Vehicles in Ten Cities" project launched in 2009. Since the beginning of the "Thirteenth Five-Year Plan," China's NEVs have moved from local demonstrations to large-scale, countrywide promotion, and the scope of central subsidies has also extended to the whole country. By 2020, it is expected that the cumulative production and sales volume of NEVs will exceed five million. In January 2016, to further ensure the smooth promotion of NEVs and cultivate a good application environment for such vehicles, the Ministry of Science and Technology issued the Notice of New Energy Vehicle Charging Facilities Incentives and Strengthening the Promotion and Application of New Energy Vehicles during the Thirteenth Five-Year Plan [33]. This notice clarified that during the 2016–2020 period, the central government will continue to arrange funds to award and subsidize NEV promotion and to construct and operate related infrastructures in various provinces (autonomous regions and municipalities). Table 2 shows three levels of China's provinces (autonomous regions and municipalities) for promotion and application of NEVs during 2015–2025. It shows the planned number of NEVs to be promoted each year from 2015 to 2025 and their proportions over the total amount of different vehicles in different levels of provinces. Between 2008 and 2017, the number of NEVs promoted in China was as follows: 500 vehicles at the 2008 Beijing Olympics; 1,300 at the 2010 Shanghai World Expo; 2011 vehicles at the 2011 Shenzhen Universiade; 27,400 vehicles in the "Thousand New Energy Vehicles in Ten Cities" demonstration and promotion project, 2009–2012; 17,643 NEVs promoted in 2013; 74,763 NEVs promoted in 2014; 383,285 vehicles promoted by 88 model cities in 2015; 507,000 vehicles promoted in 2016; and 777,000 vehicles promoted in 2017. By the end of 2017, China's total number of promoted NEVs exceeded 1.77 million, and ownership accounted for 50% of the worldwide total. Thus, China has the largest number of NEVs in the world.

At present, NEVs promoted in China are mainly pure electric vehicles and plug-in hybrid electric vehicles. However, fuel cell electric vehicles have also been steadily promoted. In early 2017, China started a new round of demonstrations for fuel cell vehicles, promoting 112 fuel cell vehicles in Beijing, Foshan, Shanghai, Yancheng, and Zhengzhou, which included 36 buses, 41 cars, 30 logistics vehicles, and 5 logistics trucks. In 2022, Yutong—the first company in China to obtain hydrogen fuel cell bus production qualification—will serve the Winter Olympics at Zhangjiakou with its fourth-generation hydrogen fuel cell buses [33].

Table 2. Planned annually increasing amount of new energy vehicles and their proportions over the total amount, including conventional vehicles, during 2016–2020 in various Chinese provinces and cities [33]

Year	Key areas		Central provinces		Other regions	
2016	30,000	2%	18,000	1.5%	10,000	≥1%
2017	35,000	3%	22,000	2%	12,000	≥1.5%
2018	43,000	4%	28,000	3%	15,000	≥2%
2019	55,000	5%	38,000	4%	20,000	≥2.5%
2020	70,000	6%	50,000	5%	30,000	≥3%

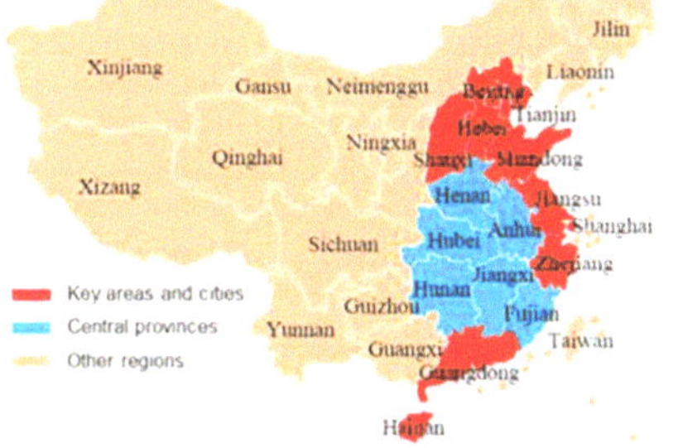

Figure 2. Three levels of China's provinces for promotion and application of NEVs during 2016–2020 [33]

3.2 Market Prediction for End-of-Life NEVs Recycling

In recent years, various favorable factors for the development of NEVs have emerged, and China's NEV industry is developing rapidly. In 2015, the sales volume of NEVs in China exceeded 300,000, accounting for 1.3% of the overall automotive industry, thus marking a turning point in the development of NEVs. In 2017, the proportion of NEVs in the automotive industry reached 2.74%. The Made in China 2025 plan states that by 2020, annual sales of Chinese brands of NEVs will exceed one million in China; by 2025, annual sales of NEVs will reach three million. Chuanfu Wang, president of BYD Auto, has predicted that NEVs will reach 30% of the total number of fossil fuel vehicles in 2025. Based on the above analysis, and according to environmental requirements, technological development trends, and historical promotion data from 2014 to 2017, this study forecasted the sales and scrapping of NEVs in China from 2020 to 2025 (Table 1). The forecast assumes that with various subsidy policies and favorable factors, coupled with continuous improvement in technical performance and continuous reductions in cost and front cardinality, the growth rate of annual sales of NEVs from 2018 to 2020 will be maintained at more than 40%. From 2021 to 2025, the growth rate will be stable at around 20%. As shown in Table 1, by 2025, sales of NEVs will exceed 5.3 million. Table 1 also shows that the number of end-of-life NEVs in China in recent years has been very small. However, by 2020, the number of such vehicles in China will reach 100,000. By 2025, the number of end-of-life NEVs in China will rapidly increase to more than 800,000.
The recycling of end-of-life NEVs in China is still in its infancy, and the amount recycled is very small. It is impossible to rely on market forces alone to establish an independent NEV recycling system. In addition, the recovery of power batteries should take place earlier than the recycling of the whole NEV. The China Automotive Technology and Research Center predicts that by 2020, the cumulative scrap of electric vehicle power batteries in China will reach 120,000–170,000 tons. A 20-gram cell phone battery can pollute 1 square kilometer of land for about 50 years; thus, several tons of NEV batteries will cause more serious environmental damage [16]. Therefore, in view of the great potential harm of scrapped power batteries and the urgency of recovery, the Chinese government should immediately work on establishing a recycling industry system for end-of-life NEVs, and especially a power battery recycling system.

4 Methodologies for Recycling End-of-Life New Energy Vehicles

4.1 Concepts of Cascade Use, Reuse, Remanufacturing, Recycling, Recovery and Residue

Cascade use is a general methodology for the efficient use of device materials and energy. The utilization of end-of-life NEVs and their parts and components can generally be divided into reusing, recycling, and recovering. According to the Technology Policy for Auto Products Recycling, issued in China in 2006, reuse refers to any use of scrapped vehicle parts for their design purposes. The reuse of parts and components retains their highest value. Some parts can be used directly while some must be remanufactured before use. Remanufacturing is necessary if the parts and components cannot be used directly; for example, an excessively worn engine shaft can be reused through remanufacturing by means of plasma spray coating technology. Recycling refers to reprocessing waste material so it can meet its original use requirements or be used for other purposes, not including the process of generating energy. Recovery refers to meeting the original use requirements or being used for other purposes after the waste material is reprocessed, including energy-generating processes. Residue refers to used material that cannot be recycled and is put in a landfill. Domestic and international end-of-life vehicle recycling industries have emphasized two indicators: recyclability rate and recoverability rate. Recyclability rate is the percentage (mass percentage) of a new vehicle that can be recycled or reused; recoverability rate is the percentage of a new vehicle that can be recovered or reused (mass percentage). Table 3 shows the relationships among reuse, recycling, recovery, recyclability rate, and recoverability rate.
China's Technology Policy for Auto Products Recycling of 2006 required that the recoverability rate of automobiles produced and sold in China be synchronized with advanced international levels starting in 2017, and the recoverability rate of all domestic and imported vehicles should reach about 95%, with a material recyclability rate of no less than 85%. Reuse retains the highest value for spare parts. Some used parts and components can be reused directly, and some can be reused after being remanufactured. To maximize use and conserve resources, end-of-life new energy vehicles and their parts are generally considered for reuse or cascade use first. Then, consideration is given to implementing remanufacturing to fully tap the added value of the raw materials, energy, labor, etc., contained in used products. Finally, the recovery of resource-based materials is considered.
China's NEV recycling industry could learn from the development history of the conventional vehicle recycling and dismantling industry over the past two decades. Moreover, it can draw lessons from the development of the conventional end-of-life vehicle industry and increase recycling efficiency of end-of-life NEVs.

Table 3. Relations of recyclability rate and recoverability rate with 4Rs (reuse, recycling, recovery, and residue)

Reuse		Recycling	Recovery		Residue
Reuse of parts	Reuse of materials	Recycling of materials	Recovery of materials	Recovery of energy	Residual materials
Recyclability rate--------------------------------------→					
Recoverability rate--→					
Total mass of vehicle --→					

Note: Reuse of parts includes different levels, such as cascade use or even reuse after remanufacturing
Source: [34]

4.2 Modes of Recycling and Reuse of NEVs

The biggest difference in dismantling technology between end-of-life NEVs and conventional vehicles lies in the battery packs in NEVs, which often have a voltage of more than 300 V [17]. If batteries are manually dismantled, accidents such as electrical shock are likely to occur. Furthermore, if a power battery short-circuits, its instantaneous current can be 100 A or more, immediately releasing a large amount of heat, which can easily cause a fire or even an explosion. As such, typical conventional vehicle recycling and dismantling enterprises do not have the ability to dismantle end-of-life NEVs. Such vehicles thus require professional recycling and dismantling companies qualified to recycle them. End-of-life new energy vehicles can only be reused after they have been safely recycled.

In January 2017, the Implementation of the Extended Producer Responsibility System, issued by the General Office of the State Council of China, required the integration of basic information regarding automobile production, trading, maintenance, insurance, and scrapping. It required gradually establishing a national unified automobile life cycle information management system and strengthening the recycling management of end-of-life automotive products. The improvement and implementation of the extended producer responsibility system plays an important role in guiding the establishment of a recycling and utilization industrial system for NEVs. Different types of NEV manufacturers, as well as the related parts and components manufacturers, have to be responsible for the recycling of their own scrapped products. The Notice on the Financial Support Policy for the Promotion and Application of New Energy Vehicles for 2016–2020 specifically states that as the main body responsible for the recovery of power batteries, NEV manufacturing enterprises are responsible for the recycling of power batteries.

There are three possible modes for establishing a recycling system for NEVs in China. The first mode is to use the conventional vehicle recycling system and simultaneously increase the recycling capacity of conventional automobile recycling companies. In the current situation, this mode is the most realistic way to recycle NEVs. For example, as a conventional vehicle recycling company, Wuhan Dongfeng Hongtai Automobile has recovered a number of NEVs. The second mode is to establish an independent recycling system for end-of-life NEVs while the government issues NEV recycling qualifications separately. However, this mode is not sufficient to support the profit of recycling companies. The third mode is to establish an independent recycling system for the key parts of NEVs and, in particular, to establish an independent recycling system for scrapped power batteries. This also includes the recovery and reuse of key components such as scrapped motors, controllers, and chargers. The scrapping of NEV bodies and other parts (tires, seats, etc.) is delegated to the conventional automobile recycling system. This mode is particularly suitable for the relevant responsible entities under the current extended producer responsibility system.

At present, the recycling and reuse of scrapped power batteries is the key issue in end-of-life NEV recycling. In 2011, China initiated power battery recycling for NEVs, set up a working group, and signed the Memorandum of Understanding on Cooperation in the Field of Power Battery Recycling with the German Ministry of Environment. Moreover, in 2014, it released the Research Report on the Feasibility of Recycling of Electric Vehicle Power Batteries in China. Since 2016, the Ministry of Industry and Information Technology of China issued, successively, the Recycling Technology Policy of Electric Vehicle Batteries (2015 edition), Industry Standard Conditions for the Comprehensive Utilization of Scrapped Power Batteries for New Energy Vehicles, and the Interim Measures for Management of the Industry Standard Conditions for the Comprehensive Utilization of Scrapped Power Batteries for New Energy Vehicles. These documents identified the main entities responsible for recycling used batteries as well as the standard conditions and management methods for the comprehensive utilization of scrapped power batteries. In January 2018, eight departments in China, including the Ministry of Industry and Information Technology, introduced the Interim Measures for the Management of Recycling of Powered Batteries for New Energy Vehicles. It aims to further strengthen the industrial management of power battery recycling for NEVs, standardize industry development, promote comprehensive resource utilization, and protect the environment. Since March 2018, pilot work on power battery recycling for NEVs has been organized and implemented nationwide. The Chinese government has therefore carried out and begun to implement top-level designs for establishing an independent industry for the comprehensive utilization of scrapped new energy vehicle batteries. China's future recycling model for NEVs thus aims to establish an independent industry model to utilize the scrapped power batteries of NEVs.

4.3 Whole-Vehicle Reuse for End-of-Life NEVs

NEVs are currently being promoted only in some large and medium-sized cities in China, and fuel cell electric vehicles are uncommon in major cities. Therefore, NEV recycling can, at present, serve as a vehicle to promote and spread knowledge about NEVs at schools, science and technology museums, exhibition halls, and so on. Currently, some NEVs that were operated in the Beijing Olympic Games, the Shanghai World Expo, and the Ten Thousand New Energy Vehicles project have been scrapped. It is recommended that NEV demonstration cities or regions should establish local exhibition halls to present these end-of-life NEVs and promote knowledge about them as well as environmental protection and energy conservation. Second, end-of-life NEVs and their key components can also be used as teaching and science materials at various schools. Only by engaging with NEVs can students fully comprehend their working principles as well as the low-carbon environmental protection they provide. Then, students may be motivated to purchase such vehicles in the future.

4.4 Recycling the Key Components of End-of-Life NEVs

Because of their fast growth, the authors identified three key components for recycling of NEVs in China:

Remanufacturing End-of-Life NEVs Parts

Remanufacturing automobile parts refers to mass producing old automobile parts using advanced methods to restore them to the same quality as the originals. The auto parts remanufacturing industry has great potential for development in China. In 2008, China promulgated the Auto Parts Remanufacturing Pilot Management Measures for the recycling and reuse of conventional auto parts. Vehicle manufacturers and auto parts remanufacturing companies were selected to carry out pilot remanufacturing, and the development of the conventional auto parts recycling industry was actively promoted. Compared to new products, the cost of a remanufactured conventional car engine is reduced by approximately 50%; energy savings are approximately 60%, and 70% is saved on raw materials. Remanufacturing is therefore very profitable for parts and components merchants. Studying remanufacturing in the automotive aftermarket, Huisingh et al. [35] noted that remanufacturing is the main link in transforming the automobile industry into one characterized by sustainable development and environmental protection. Accordingly, the China Automotive Industry Association Auto Parts Remanufacturing Branch was established in April 2010 to implement national guidelines and policies to revitalize the automobile remanufacturing industry. Remanufacturing the key components of NEVs can retain the value of the original components, which is the first choice for recycling NEVs. Remanufacturing is also an important step in establishing an independent recycling system for NEVs and their parts. Relevant research institutes and related departments should cooperate with government authorities to explore and promote remanufacturing technologies for NEVs and their parts and components. Currently, highly valued key components, such as motors and electronic controls, are suitable for reuse through remanufacturing technologies.

Cascade Use and Material Recovery of Power Batteries

Under normal circumstances, when power battery capacity is attenuated to 80% of the total capacity, the driving range drops dramatically. At that time, the power battery will need to be replaced by a new one. In addition, transformation and technological innovation also accelerate the renewal of power batteries. Therefore, power batteries are generally replaced 2–3 times during the life cycles of NEVs. After a NEV's power battery is scrapped, it is generally preferred to consider cascade use first and then consider material recovery and final disposal. Through the recovery and remanufacturing of power batteries and fuel cells, more of the value of the original parts can be retained and cost can be reduced. If these power batteries can be effectively recycled and reused after being scrapped, it can save related resources and reduce the cost of electric vehicles, which can also effectively reduce the price of NEVs and promote their sale. The cascade use of power batteries can be broadly divided into three stages. In the first stage, after testing and processing—provided the battery is intact, is undamaged, and has effective functional components—it can be used in the fields of communication and energy storage (e.g., new energy distribution power stations, street lights, communication stations). In the second stage, when the power battery is eliminated from the energy storage device or low-speed electric vehicle for the second time, there may be a third or fourth reuse, and the maximum value of reuse can be achieved. At the same time, it is necessary to make a file for each recovered battery to record its usage in detail. In the third stage, power batteries that cannot be used in a cascade are recycled, dismantled, and used for resource regeneration—that is, material recycling. Through cascade use, not only is the pollution problem solved but a good recycling model is also formed. At present, China's NEVs mainly use lithium-ion batteries. The recycling of lithium-ion batteries will become an important means to alleviate resource bottlenecks. To balance energy consumption and water pollution issues, recycling companies mainly adopt the "combination of dry and wet" method to dismantle lithium batteries and recover cobalt, nickel, and other precious metals. Scrapped lithium batteries usually contain 5–15% cobalt, 2–7% lithium, and 0.5–2.0% nickel. Further development is still needed in China's processing technologies for lithium battery recycling. The cost of recycling lithium carbonate from lithium batteries is still more than five times that of a company's direct production costs, mainly due to the lack of a unified standard for lithium batteries in China. Battery composition varies. Even with the same kind of ternary materials, the composition ratio can be very different, making it difficult to commercialize the recycling of high-added-value intermediate products (e.g., cathode materials, anode materials, electrolytes, diaphragms) from scrapped lithium batteries. China's Industry Standard Conditions for the Comprehensive Utilization of Scrapped Power Batteries for New Energy Vehicles encourages the combined use of physical and chemical methods to explore biometallurgical methods for recycling. Additionally, with wet smelting, the combined recovery rate of nickel, cobalt, and manganese should not be less than 98%. With the pyrometallurgical method, the combined recovery rate of nickel and rare earths should not be less than 97%. The Ministry of Industry and Information Technology of China has been actively guiding the promotion of pilot projects for the recycling of NEV power batteries and has obtained positive results. China Iron Tower Co., Ltd., has built five power battery cascade-use base stations in Huizhou City, Guangdong Province, and intends to demonstrate the application of energy storage, energy preparation, and peak load shifting at these base stations. Guangdong Guanghua Technology Co., Ltd., built a recycling line of 1,000 tons of end-of-life power batteries in Shantou City, Guangdong Province. Meanwhile, a recycling project involving 10,000 tons of end-of-life power batteries is also under construction in Zhuhai City, Guangdong Province. After cascade use, Tesla Electric Vehicles was able to reduce the cost of the 18,650 cylindrical batteries it used from 2007 to 2012 by about 40% through recycling and scale effects. This shows that a good recycling model can also generate huge economic benefits.

Recycling and Material Recovery of Fuel Cells

Fuel cells are composed of anodes, cathodes, electrolytes, catalysts, separators, and shells. They convert the chemical energy of the fuel into electrical energy through electrochemical reactions. The reaction materials (fuel and oxidants) are continuously consumed during electrochemical reactions to produce electrical energy and constant discharge. Currently, the number of fuel cell electric vehicles in China is very small. The total number of fuel cell sedans, buses, and sightseeing buses in the Beijing Olympics, the Shanghai World Expo, the Shenzhen Universiade, and the 'Ten Thousand New Energy Vehicles' project was only 281. Some of those fuel cell vehicles have already been scrapped. The small scale of fuel cell production and usage has made it difficult for ordinary consumers in China to access it. Therefore, to recycle fuel cell electric vehicles, they should first be used as scientific exhibits and displayed in automobile exhibition halls, science and technology museums, and various types of school laboratories to promote public awareness and science education. In addition, end-of-life fuel cells can also be used as storage power sources to provide energy for certain special cases. For example, they can be used as small independent power sources in remote areas, on islands, and in deserts. They also have broad market prospects in defense communications, combat weapon power supplies, miniature power sources, and sensor devices. Fuel cells that cannot be reused are subject to material recovery. The key materials of the membrane electrode assembly are mainly recovered, including precious metal catalysts (platinum electrocatalyst), proton exchange resins, and gas diffusion layers. Platinum cannot be artificially synthesized and is a very scarce and expensive element. Its recovery method mainly uses concentrated sulfuric acid, concentrated nitric acid, or the mixed acid of any ratio of concentrated sulfuric and concentrated nitric acid to treat the scrapped membrane electrode. Acid treatment reduces the interactions between proton exchange resin chains in the proton exchange membrane and in the catalytic layer, and it separates proton exchange resin chains from each other to disperse in acid to form a solution. The carbon carriers of the supported catalyst are oxidized, and the catalyst noble metals are fully separated.

5 Strategies for Recycling End-of-Life New Energy Vehicles

5.1 Designing and Manufacturing NEVs with Easy Dismantling Structures

Automotive recyclability and detachability designs are the basis for improving the recovery and reuse of NEVs. In the early stages of designing and manufacturing automobiles, NEV manufacturers must consider the issue of recycling and reusing scrapped vehicle parts, consider improving the utilization rates of these dismantled parts and materials, and constantly improve the process technology. Designers should fully consider the detachability of the car structure design. This includes having as few connecting points as possible; having a unified connection method where a few simple tools can disassemble most connection points; and avoiding cutting, stripping, and other projects as much as possible. Only in the initial stage of product design, taking detachability as the design goal, can we ultimately achieve efficient product recycling. At present, many NEVs have not yet taken shape and are still in the process of design and improvement. The above design principles can thus be implemented and embodied in the design of NEVs. To facilitate the recycling of scrapped automotive materials, when designing NEVs, designers should try to choose environmentally friendly materials that are easy to recycle and control the use of toxic and hazardous substances (e.g., lead, mercury, cadmium, and hexavalent chromium) while reducing the number of material types, unifying materials, and achieving unified material standardization and marking. Toyota and Ford are negotiating using the same materials to design the same parts as much as possible, showing the possibility of an international unification of automotive materials and the standardization of parts materials. These efforts aim to facilitate the classified disposal of auto parts after dismantling.

5.2 Increasing the Recoverability Rate and Recycling Price of End-of-Life Vehicles

End-of-life vehicles (ELVs) are a very important material resource and have high added value. The automotive recycling method directly determines the recoverability rate and utilization level. Remanufacturing is an advanced form of circular economy in the automotive industry that can maximize the retention of high added value in products. Therefore, we must use remanufacturing methods to recycle scrapped auto parts. The revised version of China's Management Rules of Recycling End-of-Life Vehicles is due to be released. It is understood that the revised measures will eliminate the mandatory melted destruction of the five major assemblies of ELVs and allow them to be sold to remanufacturing enterprises, thus realizing the recycling of end-of-life vehicle resources. Parts remanufacturing can save resources while greatly increasing the added value of ELVs. This helps expand the profitability of automobile dismantling companies, allows companies to recycle ELVs at higher prices, and increases private owners' enthusiasm for writing off used cars. Taking the automotive gearbox as an example, the total weight of its materials is only about 45 kilograms, and its value will not exceed 1,000 CNY if it is sold as metallurgical materials. If the part is remanufactured, its value will exceed 7,000 CNY. Therefore, the comprehensive utilization of ELVs needs to create a full-industry-chain collaboration model under which scrapped vehicles' recycling, dismantling, remanufacturing, sorting, and parts processing all take place.

5.3 Improving Regulations for NEV Recycling and Establishing an Efficient Recycling System

Although NEVs have not been marketed on a large scale globally, the recycling system for power batteries has attracted the attention of governments and enterprises in many countries. Japan has established a nickel-hydrogen-powered battery recycling system for battery production, sales, recycling and reprocessing. China is gradually establishing a complete vehicle power battery recycling system. At present, recycling NEVs still depends on the recycling system for conventional cars, which faces increasing numbers of conventional end-of-life vehicles. In addition, the recovery of power batteries, as well as fuel cells, requires special processes and technologies for recycling and reuse. Reliance on conventional vehicle recycling channels will hinder the recycling of NEVs and power batteries. Therefore, China should seek to establish an independent NEV recycling system, improve the laws and regulations for NEV recycling, and grant independent recycling qualifications for NEVs and parts to enterprises with recycling capability. Even if establishing an independent NEV recycling company is feasible, conventional vehicle recycling companies will need to establish independent NEV recycling departments as separate businesses. Departments of automobile management, quality supervision, industry and commerce, and environmental protection can strengthen legal supervision and control and effectively increase the actual recycling rate of new energy automotive products in China. Relevant scientific research institutes and related departments should cooperate with government authorities to explore policies, management systems, and regulatory systems that will promote the development of NEV parts recycling and remanufacturing industries. Moreover, they should study domestic NEV parts trading and remanufactured product sales to adjust relevant management policies and accumulate experience to establish technical standards for remanufacturing, market access conditions, and circulation supervision systems.

5.4 Improving the Exit Mechanism of End-of-Life Vehicles in China

To promote the orderly development of the dismantling industry for end-of-life vehicles, it is recommended to first establish reasonable policy guidance and industrial planning, and improve the exit mechanism for end-of-life vehicles. The current exit mechanism for end-of-life vehicles is incomplete. In particular, the implementation of mandatory vehicle scrapping is not effective. An owner is allowed to purchase a new vehicle without having his or her end-of-life vehicle recycled. Furthermore, illegal dismantling and underground transactions are still rampant, which means that many (quasi) end-of-life vehicles more likely go to the illegal dismantling market than to scrapping. To solve the problem of where end-of-life vehicles go, the key is to establish a linkage mechanism, strengthen coordination between departments of commerce, public security, environmental protection, and transportation, and open up the blockage points in the whole life cycle of automobile production, circulation, and scrapping. It is necessary to strengthen market supervision and crack down on illegal dismantling and black-market transactions; on the other hand, vehicle owners need to be actively guided to go to official vehicle recycling companies to scrap their vehicles. Owners should be made aware that if their vehicles are sold to the illegal dismantling market, the license plate cannot be canceled and the vehicle will still be attached to their name. In the event of a traffic accident or other violation, the original owner would therefore have to take responsibility. Therefore, it is equally important to increase awareness of safety risks among vehicle owners while also strengthening crackdowns on illegal operations.

5.5 Strictly Implementing the Extended Producer Responsibility Policy

Since the manufacturer's recycling responsibility policy lacks needed improvements and enforcement, most Chinese automobile manufacturers are only responsible for the production of cars, and they are less concerned with recycling. At the beginning of 2017, the General Office of the State Council issued the Implementation of Extended Producer Responsibility Policy and proposed that electric vehicle and power battery manufacturing enterprises should be responsible for establishing a waste recycling network. In March 2018, seven ministries and commissions jointly issued the Interim Measures for the Management of Recycling of Powered Storage Batteries for New Energy Vehicles, emphasizing the need to implement the extended producer responsibility policy. Under the requirements regarding who produces and who is responsible, vehicle manufacturers assume the main responsibility for power battery recycling, and related production companies fulfill the corresponding responsibilities in each step of recycling. It is very convenient for vehicle manufacturers to recycle. They are more familiar with the parts they produce, can more easily dismantle vehicles, and have inherent advantages in parts remanufacturing. Component manufacturers and raw material producers are responsible for recycling their own parts or raw materials. They can provide technical support, provide the relevant testing and dismantling required by the process of raw material development and parts composition for the related resource recycling companies, and provide financial support. For raw materials recovered by new energy auto parts and resource recycling companies, the corresponding raw material manufacturers and parts manufacturers can fulfill the purchases of a certain ratio of recovered materials. In the recycling of end-of-life NEVs, relevant government departments, NEV manufacturers, and recycling companies need to take measures to build a sound new energy vehicle and parts recycling system. For government departments, they can first establish sound policies, appropriate regulations, and a good development environment for the automotive recycling industry. Second, they need to strictly implement the producer responsibility system, establish a recycling database, and incorporate recoverability rate into the new product certification system of manufacturing companies. Finally, they can use various economic and taxation levers to increase the competitiveness of auto recycling companies and promote their healthy development. Automobile manufacturers need to make great efforts to develop green designs and manufacturing technologies that increase the recoverability rate. Recycling companies can improve vehicle dismantling and remanufacturing technologies. China's end-of-life NEV recycling industry will only experience continuous development with the implementation of a sound recycling system and constant improvements in production and recycling technologies.

6 Conclusions

Based on the current status and problems associated with the recycling and reuse of traditional end-of-life vehicles in China, the present study has investigated the current state of NEVs' market promotion and forecasted their production and sales in China. Furthermore, this paper analyzed in detail the models and the specific recovery and reuse methods for reclaiming and reusing upcoming end-of-life NEVs and their key components. Finally, a number of marketing strategies have been proposed to promote the recycling and reuse of end-of-life NEVs. Based on the above analysis, this paper draws the following conclusions: (1) The system for constructing and recycling new energy vehicles should be planned in advance. (2) China's future recycling model for NEVs should establish an independent industrial model for the comprehensive utilization of scrapped power batteries for NEVs. (3) For the recycling of end-of-life NEVs and their key components, the government, as well as NEV production and recycling enterprises, must take action to create a sound NEV and parts recycling system. Implementing the producer responsibility system is key. (4) The recovery and reuse of power batteries is the core aspect of the recycling of end-of-life NEVs. Thus, it is necessary to focus on improving power battery recycling policies and systems.

7 Zusammenfassung

China hat weltweit die meisten „New Energy Vehicles" (NEV). Im Laufe der Zeit wird ein Teil dieser NEV jedes Jahr ausrangiert werden. Dieser Beitrag untersucht den potenziellen Markt für NEV am Ende ihres Lebenszyklus sowie bestehende Probleme in den Wieder- und Weiterverwertungssystemen Chinas, indem Status Quo und Herausforderungen beim Recycling konventioneller Fahrzeuge betrachtet werden. Es zeigte. sich, dass die Anforderungen an das Recycling- und Weiterverwertungssystem frühzeitig geplant werden sollten. Zudem wurden verschiedene Verfahren zum Recycling und zur Wiederverwendung von NEV und deren Komponenten analysiert und darauf aufbauende Strategien entwickelt. China sollte als zukünftiges Recyclingsystem für NEV das Modell einer unabhängigen Industrie wählen, um die in NEV enthaltenen Batterien umfassend zu nutzen. Die durchgeführte Studie stellt einen Referenzpunkt für ein robustes NEV-Recyclingsystem sowie zur Umsetzung einer Recycling- und Wiederverwendungsplattform dar.

Acknowledgments. We appreciate the supports from the IPID4all Senior and Student Research Exchange projects of the German Academic Exchange Service (DAAD) with funds from the Federal Ministry of Education and Research (BMBF). Thanks also to National Natural Science Foundation of China (No. 51476120) and the 111 Project of China (No. B17034).

8 References

[1] Li YY, Wang LH (2016). The recovery status, forecast and countermeasures for end-of-life vehicles of China. Ecological Economy, Vol.32, No.6, pp 152-156. doi: 10.3969/j.issn.1671-4407.2016.06.031

[2] He ZG, Ye LP, Liao W (2012). Key issues in development of recycling and dismantling enterprises of end-of-life vehicles in China. Recyclable Resources and Circular Economy, Vol.5, No.9, pp 32-36. doi: 10.3969/j.issn.1674-0912.2012.09.015
[3] Fang ZX (2014). Overview of policies and standards for China's scrap automobile recycling and parts remanufacturing. Automobile Parts, No.1, pp 81-85. doi: 10.19466/j.cnki.1674-1986.2014.01.025
[4] Zhang H, Chen M (2014). Current recycling regulations and technologies for the typical plastic components of end-of-life passenger vehicles: a meaningful lesson for China. Journal of Material Cycles & Waste Management, Vol.16, No.2, pp 187-200. doi: 10.1007/s10163-013-0180-3
[5] Wang L, Chen M (2013). Policies and perspective on end-of-life vehicles in China. Journal of Cleaner Production, Vol.44, pp 168-176. doi: 10.1016/j.jclepro.2012.11.036
[6] Xiang W, Chen M (2011). Implementing extended producer responsibility: vehicle remanufacturing in China. Journal of Cleaner Production, Vol.19, pp 680-686. doi: 10.1016/j.jclepro.2010.11.016
[7] Shen J (2009). Investigation on the passenger vehicles recovery system in China: on the principle of extend producer responsibility. Shanghai Jiaotong University, Shanghai.
[8] Fang HF (2009). Push auto recycling by manufacturer extended responsibility (II). Automobile & Parts, ISSN: 1006-0162. No.32, pp 43-36. http://www.cqvip.com/read/read.aspx?id=31237870
[9] Li YK, Gao Y (2013). Experiences and inspiration of traction battery recycling in Germany. Renewable resources, ISSN: 1673-7776. No.10, pp 48-50. http://www.cqvip.com/read/read.aspx?id=47669155
[10] Hu T, Cao CM, Wu YP, et al (2010). Experience on recycling and disassembling of end of life vehicle in Japan and its implications for China. WTO Economic Herald, No.10, pp 74-77. ISSN: 1672-1160. doi:10.3969/j.issn.1672-1160.2010.10.017
[11] Zhao Q, Chen M (2011). A comparison of ELV recycling system in China and Japan and China's strategies. Resources Conservation & Recycling, Vol.57, pp 15-21. doi: 10.1016/j.resconrec.2011.09.010
[12] Xia XH, Wan G (2018). Optimistic about the sales of new energy vehicles to 1 million cars this year. Electric and mechanical business, 2018-01-29 (A01). http://www.meb.com.cn/news/2018_01/30/6202.shtml
[13] Wan G (2018). New challenges and new directions of new energy automotive industry. Automotive Longitudinal, ISSN: 2095-1892.No.02, pp 18-20.
[14] Miao W (2018): The development of new energy vehicles in China is facing five major problems. http://www.ce.cn/xwzx/gnsz/gdxw/201801/21/t20180121_27824568.shtml
[15] Han W, Zhang G, Xiao JS, et al (2014). Demonstrations and marketing strategies of hydrogen fuel cell vehicles in China. International Journal of Hydrogen Energy, Vol.39, pp 13859-13872. doi: 10.1016/j.ijhydene.2014.04.138
[16] Ding X (2017). Analysis of the development of China's new energy vehicle scrapping and dismantling industry. China Resources Comprehensive Utilization, Vol.35, No.12, pp 93-95. doi:10.3969/j.issn.1008-9500.2017.12.033
[17] Yu HJ, Xie YH; Ou YN, et al (2014). Industry exploration dismantling and recycling of new energy vehicle in our country. Environmental Science and Technology, No.3, pp 66-69. doi:10.3969/j.issn.1674-4829.2014.03.018
[18] Regenerative Resources Association (2016). In 2020, the battery scrap rate of electric vehicles will reach 120 thousand ~ 170 thousand tons. China Resources Comprehensive Utilization, Vol.34, No.02, pp 43-43. doi:10.3969/j.issn.1008-9500.2016.02.020
[19] Li YK, Miao G, Ao Y (2014). Research on recycling economy of vehicle power battery. Automobile & Parts, ISSN: 1006-0162. No.24, pp 48-51. http://qikan.cqvip.com/article/read.aspx?id=81678074504849525052485153&from=article_detail
[20] Elwert T, Goldmann D, Römer F, et al (2015). Current developments and challenges in the recycling of key components of (hybrid) electric vehicles. Recycling, Vol.1, No.1, pp 25-60. doi: 10.3390/recycling1010025
[21] Wittstock R, Pehlken A, Wark M (2016). Challenges in automotive fuel cells. Recycling. Vol.1, No.3, pp 343-364. doi: 10.3390/recycling1030343
[22] Ramoni M O, Zhang H C (2013). End-of-life (EOL) issues and options for electric vehicle batteries. Clean Technologies & Environmental Policy, Vol.15, No.6, pp 881-891. doi: 10.1007/s10098-013-0588-4
[23] Naor M, Bernardes E S, Druehl C T, et al(2015). Overcoming barriers to adoption of environmentally-friendly innovations through design and strategy. International Journal of Operations & Production Management, Vol.35, No.1, pp 26-59. doi: 10.1108/IJOPM-06-2012-0220
[24] Li YK, Zhou W, Huang YH (2012). The idea of establishment new energy automotive battery recycling system. Renewable resources, ISSN: 1673-7776.No.1, pp 28-30.
[25] Lv ZY, Ma HX (2016). Design of waste battery recovery system of new energy electric vehicle. Automobile Applied Technology, ISSN: 1671-7988.No.11, pp 18-19.
[26] Hou B (2015). Research on recovery mode of electric vehicle battery. Chongqing University of Technology, Chongqing, China.
[27] Peng JL (2017). Scrap power battery recycling pretreatment plan and technology. Hefei University of Technology, Hefei, China
[28] Chen JW (2007). Analysis of the PEMFC recovery. Power technology, No.10, pp 827-829. doi:10.3969/j.issn.1002-087X.2007.10.019
[29] Analysis of China's vehicle occupancy and forecast of development trends in 2016 - Industry Trends - China Industrial Development Research Network. http://www.chinaidr.com/tradenews/2016-07/98829.html
[30] The new energy car posture gratifying core technology is still insufficient. Sohu.com 2018 [2018-03-07]. http://www.sohu.com/a/225030883_100125118
[31] http://www.chyxx.com/research/201802/612314.html
[32] http://www.chyxx.com/industry/201805/636567.html
[33] On the "13th Five-year Plan" new energy vehicle charging infrastructure reward policy and strengthening the promotion and application of New energy vehicles. http://www.most.gov.cn/tztg/201601/t20160120_123772.htm
[34] http://www.ndrc.gov.cn/zcfb/zcfbgg/200602/t20060214_59502.html
[35] Subramoniam R, Huisingh D, Chinnam R B (2009). Remanufacturing for the automotive aftermarket-strategic factors: literature review and future research needs. Journal of Cleaner Production, Vol.17, No.13, pp 1163-1174. doi: 10.1016/j.jclepro.2009.03.004

Assessment of Reusability of Used Car Part Components with Support of Decision Tool RAUPE

Alexandra Pehlken[1], Björn Koch[1], Matthias Kalverkamp[1]
[1]Cascade Use Research Group, Carl von Ossietzky University of Oldenburg, Oldenburg, 26129, Germany
alexandra.pehlken@uol.de

Abstract

Owners of a car usually have to maintain their car including their car parts regularly, which results in exchanging car parts from time to time. Normally, products made by original equipment manufacturers are used by professional garages because of the guarantee they have to provide. The newly developed decision tool RAUPE is based on data mining methods applied to a data base provided by a car dismantling network that includes around one million spare parts' data. The results of this data mining is included in RAUPE giving information on the reusability chances of the spare parts and the life cycle performance related to its CO_2 emissions. The decision tool is in its beta version and is to be evaluated by the end of 2018. The paper describes the concept of RAUPE and the role of its stakeholders.

1 Introduction

Material management is highly important in car manufacturing since most materials are derived from mining operations. Depending on the original equipment manufacturer's (OEM) location, they might be heavily dependent on importing raw materials. The supply chain for car manufacturing has already raised a lot of attention since the materials in the car are becoming more and more precious. For example, the valuable and sometimes rare elements Cobalt and Gallium or Neodymium and Dysprosium, among others, can be found in today's cars, but especially in the electric cars of the future. It remains questionable, however, where – in which components and material compounds – and in which quantities these materials can they be found and how they can be identified and recovered at the end of the useful life of a car to be redirected into the economic cycle. Currently, energy efficiency and energy carriers are dominating vehicle development and production, but the industry will have to deal much more intensively with the use of new and sometimes scarce and environmentally problematic materials and life-cycle management. In the future, more cars are supposed to be running on different powertrain concepts, as for example electric motors, and the objective is to further enhance their security, efficiency and intelligence. In doing so, they will increasingly contain electrical and electronic components, such as motors, sensors, management systems or batteries, and consequently an increased share of new materials and material combinations. Many of the resources used are rare, critical and valuable, while their extraction is entails high efforts and environmental impacts. Numerous of those materials belong, for example, to the group of Rare Earth Elements (REE), most of which are classified as strategic resources. With regards to a sustainable and economic management, they all require the best possible recovery and reuse. There are still many untapped ecological and economic potentials in this field. However, researchers and recyclers are often still lacking the necessary information regarding the amount and application of these materials, as well as the technologies that allow for their processing and reuse. Therefore, it is necessary to work on information and decision tools to overcome the lack of information outside the manufacturer business. Due to the high quality manufacturing process and the high volume of used car parts, a second hand market has been established nearly worldwide. While remanufacturing is often driven by manufacturers who reverse their supply chain to refurbish only their own products, an independent market on used car parts focuses on all brands in use in the market. This market is especially useful for often price-concerned regular car owners, as used car parts are often cheaper than parts from Original Equipment Manufacturers. A disadvantage as of today, however, is the fact that hardly any guarantee is provided for the used car parts. Currently, OEMs and garages do not exchange much information on car parts and therefore garages need to gain their own experiences by testing for example car part A in car C. Neither national nor international data is being shared.

This paper addresses the concept and prototype of a decision support tool "Recycling of Automotive Units and Parts Evaluator" (RAUPE) addressing the further usage of used car parts and their life time extension. The overall goal of RAUPE is twofold: 1) RAUPE indicates a possible reduction of the CO_2 emissions of automobiles through component reuse and 2) the potential for critical material conservation when assessing the state of automobile car parts through relevant information to support decisions.

When considering a functional automobile during its use phase, a user generally may wish to assess the environmental impact of recycling or reusing the components of the automobile against the resale price, if such a price is feasible. When assessing the automobile, questions regarding the overall mass and price of critical materials in various components or in the total automobile are raised together with questions about the resale price of components. Materials are therefore of high interest for the recycler and dismantler whereas the resale price is important for both car owner and buyer, as well as the recycler. The decision tool should thus support recycling- and consumer-related stakeholders in assessing the environmental worth and quality of their automobile with respect to its components, material CO_2 load as cumulative

A. Pehlken et al. (Eds.), *Cascade Use in Technologies 2018*,
https://doi.org/10.1007/978-3-662-57886-5_10

energy demand, and finally the reuse potential of an automobile when components fail. The decision tool is publically available and non-proprietary, only open-source technologies have been used in the development process.
The decision tool is thus composed of different levels of granularity at a car, component and material level by connecting various unconnected data sources to provide comprehensive information. This paper explains the relevant stakeholders, system boundaries, data sources and the theoretical architecture that combines all the individual parts.
Various stakeholders are involved in the development of the architecture for the decision tool, such as industry partners during implementation and public, governmental, recycling and disassembly audiences during evaluation. The industry stakeholders identified at this point of development are, in the first step, car dismantlers who have provided an index of disassembled parts for identification of car assembly, and in the second step software companies who have expressed an interest in participating in the development of the decision tool and later adoption for continuity after the lifetime of the research project Cascade Use. Having companies involved increases the impact of the decision tool by having a larger data set and longer maintenance lifetime. Possible target groups include public or governmental consumers and reviewers that have an interest in the CO_2 output of the automobile during the production, use and especially reuse and recycling phases or in the critical material content of an automobile or subset of its components. Since the production process itself is excluded here for data protection reasons, only the energy needed to extract and process raw materials needed is integrated as cumulative energy demand.

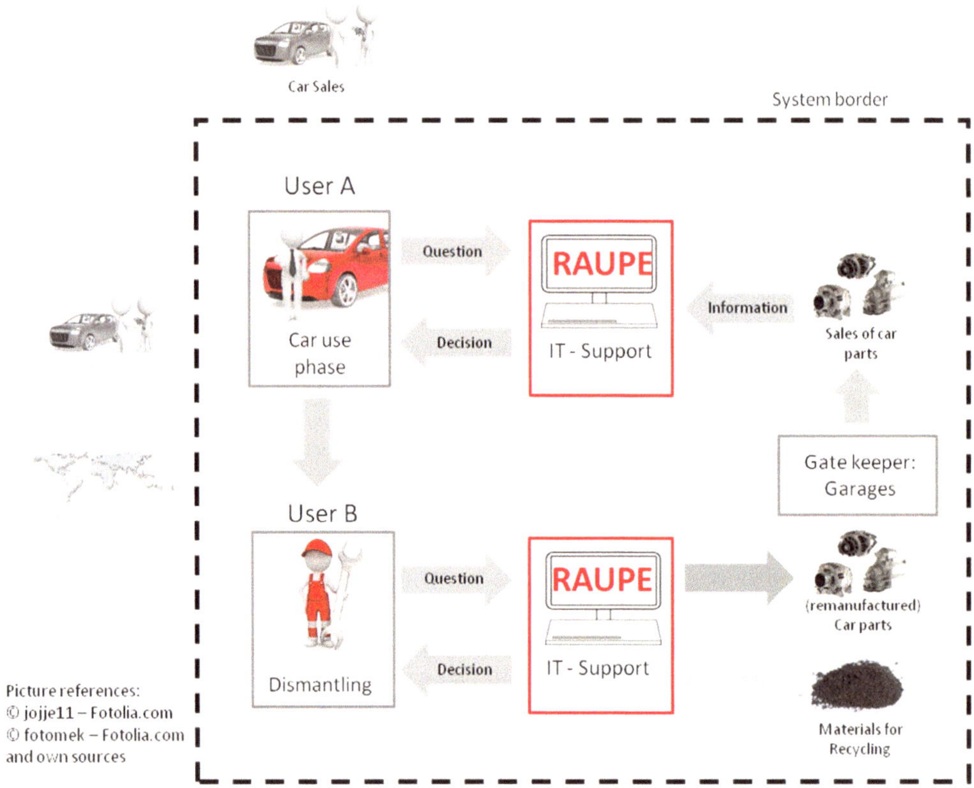

Figure 1. System Border of RAUPE

The use phase is the most addressed life cycle phase of a car and therefore RAUPE integrates data already published by other research projects, e.g. "NextGreenCar" [1]. However, the use phase strongly depends on the user behavior and is therefore highly variable according to the car owner/driver. Target groups of RAUPE users are outlined in Table 1, including their objectives and relevant information requirements. The recycler and public consumer stakeholder groups have been identified preliminarily as the evaluation groups for the decision tool when development has been completed.

Table 1. Stakeholder overview of the CCU decision tool "RAUPE"

	Objectives	Information Requirements
Public Consumers	Assessing component value Understanding environmental impact	Hazardous contents Material market values Environmental impact
Governmental Reviewers	Assessing market impact and scarcity of materials Understanding environmental impact	Material criticality Material market values Environmental impact
Disassembly Plants	Deciding whether to purchase components Estimating disassembled value	Component assembly Bill-of-materials Overall value
Recyclers	Input sorting for safety and value Recovering marketable materials	Hazardous contents Component assembly Material market values Material volume

The system boundaries of the proposed architecture must be clearly defined for the implementation to be realized in a reasonable time frame. The first main boundary is to consider which vehicles provide the basis of the database, since not every single type of car can be included. By restricting the subset to combustion, electric, battery and hybrid automobiles and excluding motorcycles, larger and specialized construction or logistics vehicles and purpose-specific vehicles, the decision tool can be developed with an adequate supply of data. The architecture designed to develop into the decision tool has the potential to support each mentioned vehicle with sufficient alterations and data availability. The second boundary to consider is how much data is reasonable to provide, with the electronic automobile components being selected as the most important case study for recycling and reuse in the decision tool. The restriction of components can be removed in the architecture, provided that sufficient data is made available by original equipment manufacturers, noting that legislation or incentives would be required to procure such data, which falls outside the scope of this study.

1.1 Legal Situation

As the case study is performed in Germany with cars that enter the German recycling chain, this chapter explains the German regulation.

The basis for the waste management regulations in Germany is the Waste Framework Directive (2008/98/EC), which was adopted on the European level. All member states of the European Union have the obligation to implement the Waste Framework Directive on the national level. Germany's Waste Management Act (KrWG) is the national law that transposes the Waste Framework Directive in the German law. In addition to the general Waste Management Act, Germany has released special regulations for specific types of product waste, for example for End-of-Life Vehicles (ELV). The waste management regulation for ELVs is governed in the AltfahrzeugV, which was released in 1997 and last updated in 2016. Since 2015, all newly registered cars have to reach the following targets at the end of their lifetime: recycling and re-use of 85% based on the car weight, and 95% for re-use as a whole [2]. These targets are complemented by facility rates. These require the dismantling facilities to re-use or recover material of at least 10% from non-metallic elements and the shredding facilities to recover material of 5% and an additional 10% of non-metallic shredding residuals. This shows the increasing importance of the reuse of car parts as spare parts for still functioning cars.

All manufacturers are legally obliged to take back their cars free of charge. They also have to ensure that the dismantling facilities are widely available, meaning within a range of 50 km in Germany [3].

The EU Directive also requires that all dismantling facilities must have a valid certification to be allowed to dismantle the cars and circulate used spare parts. This also includes the shredding facilities, which need to be certified in order to be able to operate. This aims at ensuring that environmental hazard is avoided and recycling improved.

The car manufacturers are also obliged to provide all required information on the car components to the dismantling and shredding facilities. This information includes the construction of the components, the liquids, optimization of the re-use and recycling as well as energetic recycling methods [4].

The quality of the car parts that are dismantled for reuse is determined in the VDI 8040 (Quality of recycled car parts), which is provided by the Association of German Engineers (VDI, Verein deutscher Ingenieure). The VDI 8040 is a binding regulation defining the minimum quality criteria for the German automotive used spare part market. This regulation defines the tools for the quality assessment as well as the test instructions and documentation. All used spare parts can be classified according to three different quality ranges based on component specific pre-defined quality criteria (VDI 8040 2009). This ensures that safety as well as quality standards of used spare parts are adhered to. Nevertheless, adherence to the quality criteria is more or less determined by optical assessment, which does not include any measurement or statistical assessment.

1.2 Reuse Market of Car Parts

Driven by the European Commission, Germany is also affected by the action plan on how to support the transition to a circular economy, published in 2015. The concept of the circular economy prescribes using products, materials, components or parts for as long as possible within the economy. This prolongation of the lifetime minimizes waste and reduces the environmental impact that occurs during the production process [5].
Especially, the steps re-use and recycling, which are in the middle of the waste hierarchy, are fundamental to the circular economy [6]. In an ideal world, a material, part or component would be re-used several times before being recycled several times as well. Keeping resources within the economy this way is also referred to as cascade use.
A key reason for prioritizing both re-use and recycling is minimizing the consumption of primary resources, as these are finite and thus often critical and secondly, the environmental impact that is created by mining primary resources is tremendous [7]. However, the proportion of so-called secondary resources in the market is still low, which is mainly due to uncertain quality standards [5]. In addition, there are also a economic constraints for re-use and recycling. As the complexity of products increases, innovative and often expensive recycling technologies have to be applied to achieve the targeted standard [7].
The automotive industry is one of the leading industries in re-use and recycling. This is mainly due to environmental rules and regulations as well as the very high resource potential in scrapped cars. The weight of an average car is made up of more than 10,000 different parts made of about 40 different materials which are spread in 39% bodywork, 25% chassis, 15% fitting-out, 15% drives and 6% electronics. This translates in a material composition of ~64% conventional steel, ~10% medium- to high-strength steel, ~8% polymers and composites, ~7% aluminum, ~1% magnesia and ~10% other materials [8].
In Germany, about 90% of the car weight is currently re-used or recycled. The share of reuse is still comparably low at about 4%. This is due to the fact that first, not all parts can be re-used and second, that recycling is competing for recovered parts that could also be re-used. This is mainly driven, on the one hand, by the economical optimum of dismantling versus shredding and on the other hand by the market price and demand for used car spare parts [9]. However, in the long run shifts from reuse towards recycling are not expected as sales of parts for re-use are very attractive [10].
At about 69%, recycling after shredding accounts for the highest share of the amounts recycled, followed by recycling after de-pollution and dismantling at about 10% and recycling after export at about 6% [11]. The conventional as well as medium- to high-strength steels, which make up the biggest share of material in a car, have a very high recycling rate ranging from 98% up to 100%. Aluminum, which also makes up a significant share in the car, has a slightly lower recycling rate of about 95%. These metals are usually shredded, in some cases processed further and then melted into recyclable metal. The biggest challenge for the recycling of scraped cars today is the handling of plastic components. Only 5-10% of these can be recycled, the remainder has to be used energetically or landfilled. Similar challenges occur with the recycling of glass [2, 12]. Over the next 20 years, the recycling of electronics will become more important as the share of electronics in cars has been progressively increasing over the last years. However, there are already indications that a positive recycling rate will be achieved in the future to ensure reuse and recycling targets [9]. Similar trends can also be observed e.g. in North America and China, but will not be further discussed here.

2 Architecture of Decision Tool RAUPE

In order to retrieve spare parts from ELVs the cars need to be dismantled. There are currently around 100 dismantling or car recycling stations in Germany which dismantle the ELVs according to the specified regulations. Only 2/3 of these dismantling stations obtain used spare parts from the EVLs. The remaining 1/3 focuses on recycling only [10]. The main reason that not every dismantler retrieves used spare parts lies in the difficulty to guarantee that the used spare parts can operate properly in a new car.
Here the decision tool can support in making the decision whether a part may be reused or not. The basis for this assessment is the history of more than 1 million used spare parts (from 2010 to 2017) and, respectively, the spare parts that are used while the decision tool operates, as it has a user-generated content interface.
Assembling data for the three relevant levels of detail of an automobile – material, component and automobile – requires an appropriate information model to translate the concepts into a decision tool. The architecture identified for the decision tool is the classical "three tier" architecture design which supports an abstraction of three core aspects of the decision tool. As the decision tool is designed as an interactive and responsive web based application the tiers are:

- website (presentation tier)
- RESTful application server (logic tier)
- back-end database and data store (data tier).

These three tiers or layers within the architecture are shown in Figure 2.

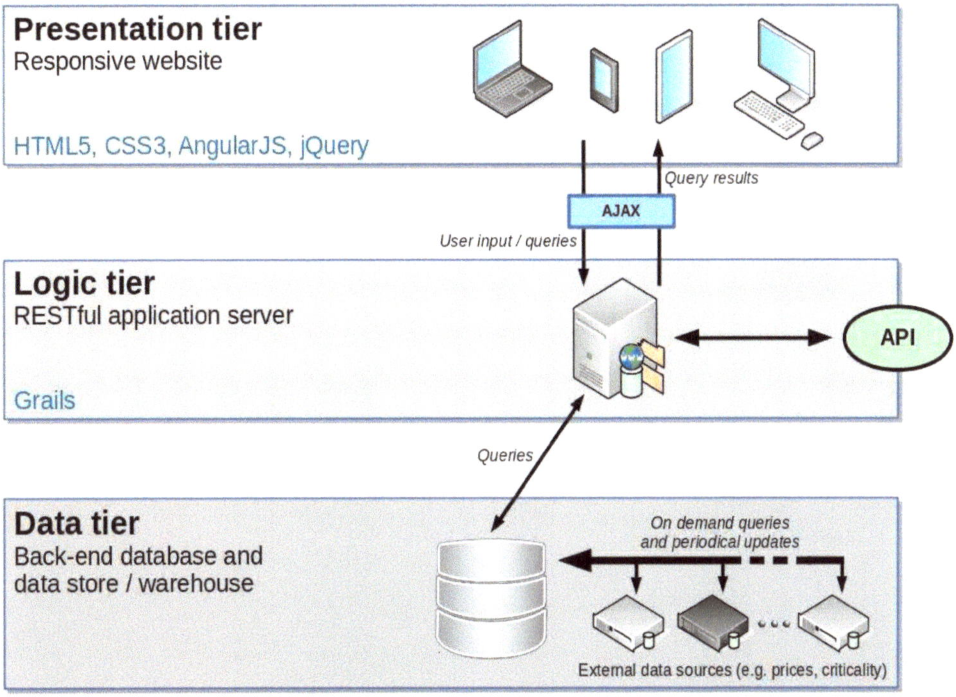

Figure 2. Three-tier architecture design of the CCU decision tool

2.1 Presentation Tier

The presentation tier is the topmost layer of the application and the user interface. It is the layer where the user feeds the decision tool with data and queries and the results are presented. In the decision tool, this layer is implemented as a responsive website using the current web standards HMTL5 and CSS3 as well as the JavaScript-based open-source front-end web application framework AngularJS and the free and open-source JavaScript library jQuery. Because of its responsive design the website will render well on a variety of devices and window or screen sizes, including mobile devices, tablets and PCs.

Depending on the stakeholder and their interests, different data blocks can be presented in the user interface, including a list of online sources of supply for a spare part that entails not only the current prices but also the impact on the environment by indicating the amount of CO_2 or raw materials saved. An additional feature in the user interface is the integration of user-generated data (UGC). Any user logged in at RAUPE has the possibility to add data and comments about spare parts to the database. This way the amount and quality of data within the decision tool will improve.

A screenshot of an early beta version of RAUPE's presentation tier can be seen in Figure 3.

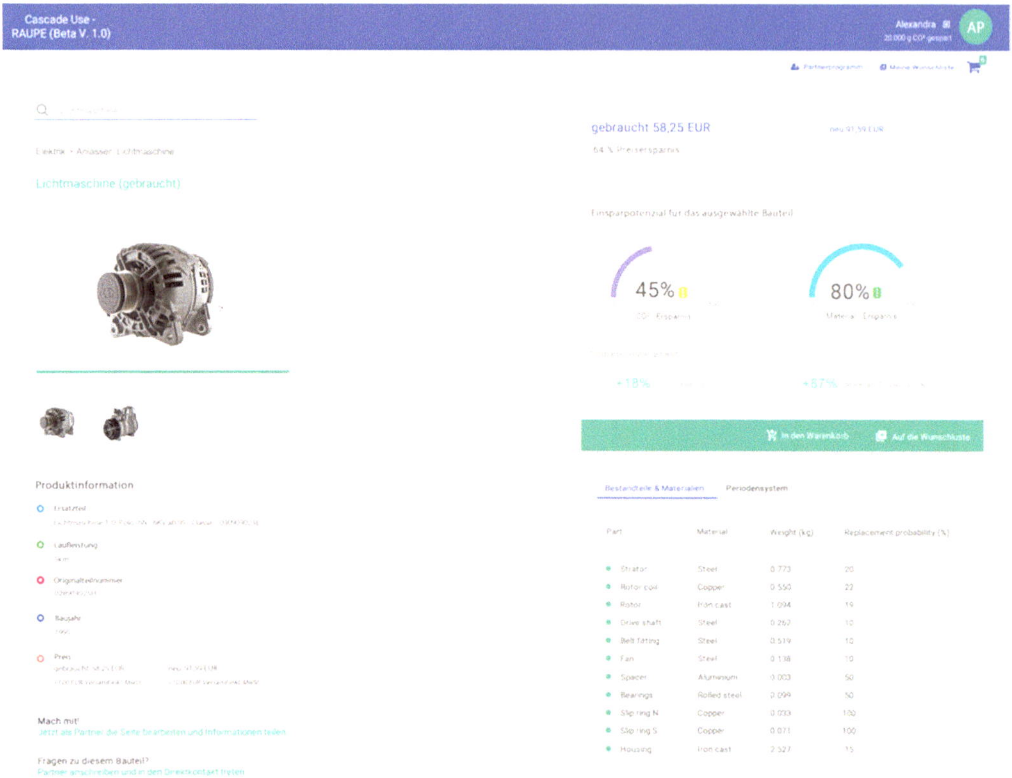

Figure 3. Screenshot of RAUPE (early beta version)

2.2 Logic Tier

The logic tier or the decision logic layer controls the application‘s functionality. On the one hand, it processes the queries and other data entered by the user in the forms of the website. Based on this data, it consults the mapping engine of the data access layer to map the user's queries with the databases of the data access layer. On the other hand, it processes all the data collected from the data access layer with the queries made by the user and sends the results back to the presentation layer which will update the website's content accordingly and display these results in a user-friendly way.

The logic tier is designed as a RESTful application server implemented in "Grails" which is an open source web application framework that uses the Apache Groovy programming language (which is in turn based on the Java platform). The entire communication between the presentation layer and the logic layer is done in Ajax (Asynchronous JavaScript and XML) but the logic tier offers an additional API (Application Programming Interface), too. As the presentation tier is completely seperated from the others, alternative interfaces can be developed either using AJAX or the API, such as the OpenLCA application of GreenDelta GmbH or even highly specialised apps based on the stakeholders' demands.

Because of the limited amount of data available about components and automobiles, fuzzy classification schemes are required to drill down from an abstract automobile to related automobiles and finally to the exact automobile queried, with a similar process followed for components to provide information in cases where data is only partially available. To improve the amount and quality of data the decision tool allows the user to submit user-generated content.

2.3 Data Tier

The data tier or the back-end database and data store is responsible for retrieving data from the local database as well as populating and maintaining the local database through web services or other external databases. The local database is the heart of the data tier and working like a cache or a data warehouse. Every query's result is stored with a timestamp. Depending on the current user's queries and the actuality of the data stored in the local database further external sources, such as external databases or web services, are queried to gather further and newer information about every kind of data processed in RAUPE, such as prices, criticality etc. ("on demand" queries). Furthermore, recurring automated updates of data are launched depending on how often information might change in the external sources (periodical updates). E.g., prices might be updated on a daily basis regardless of the user queries of a day, while information about criticality might only be updated once or twice a year as this data does not change more often. This way, the local database is kept current by invoking each service at specific intervals. The main reason for choosing a local database rather than real-time querying is the prohibitively large number of service queries required for each user request, estimated to reach too many polls to each material service for queries of the material contents of automobiles. This would cause the services to either be slow and unresponsive, or in the worst case not work at all. Additionally, external queries and so external internet traffic is reduced to a reasonable minimum, thus also reducing energy consumption and CO_2 emissions.

2.4 Database Used and Implementation at Car Dismantlers

The concept and implementation of RAUPE at car dismantlers and garages will led to more sustainable usage of car parts. Starting with a data base on used car parts sales over the last years from 2010 to 2017 and the user satisfaction report, provided by the car dismantler network "callparts", a few million data sets are already included. It must be noted that the data was received from the past and from cars that either crashed or were old enough for recycling and therefore entered the recycling route. Since most car parts are subject to maintenance operations, the used car parts itself are not necessarily as old as the car as they have been replaced two or three times during the life time of the car. A good example for this is given by passenger tires, which are normally replaced every three years during the car's life time.

3 Conclusion

The motivation displayed by the industrial partners has shown that there is a big gap between OEM data and the data that recyclers can access. Of course, information is given on how to dismantle the car, but that is nearly all. No bill of materials is being shared and therefore recyclers have to start right at the beginning in a trial and error kind of fashion. Usually, used spare parts are sold as "a pig in a poke" and if the buyer complains they get a new one without discussion. Since this kind of information is recorded in the digital scrap yard of the callparts network, the idea of RAUPE was to retrieve useful information from it. It is therefore the first attempt to perform data mining with a digital scrap yard on used car parts.

As of 2018, RAUPE is under evaluation by the car dismantlers within the callparts network and not yet mature. In addition, not every car part is included in the data base yet, since the research group was not able to perform more than 100 life cycle assessments because of time and resource constraints. RAUPE could finally be used by professional dismantling and repair shops, in order to utilize even used spare parts for professional repairs that come with a guarantee. The industrial partners have therefore shown a high interest in using this tool and are open for sharing data. In addition, it should be noted that uncertainties concerning the parts still remain since it cannot be determined exactly if the spare part was the original equipped one or a two or three times replaced one. More field studies need to be done. Nevertheless, it would be of big help if car manufacturers share more data on car parts and materials. This would dramatically increase the feasibility of the tool and as a direct result would facilitate the circular economy approach.

4 Zusammenfassung

Der vorliegende Artikel handelt von der Entwicklung eines Entscheidungstools "Recycling of Automotive Units and Parts Evaluator" (RAUPE) von der Nachwuchsforschergruppe Cascade Use der Universität Oldenburg. Sobald ein Automobil nicht mehr wirtschaftlich reparierbar ist und entsorgt wird, gehen potentiell wertvolle und noch funktionsfähige Bauteile verloren. Alternativ können sie wieder in einen neuen „Reuse"-Nutzungszyklus eingebracht werden. Autoverwerter übernehmen an dieser Stelle die Funktion, möglichst viele Produktkomponenten bzw. Autoteile zu separieren und einem Gebrauchtteilemarkt zur Verfügung zu stellen, bevor es zur stofflichen Verwertung kommt. Hieran ist besonders, dass Verwerter meist Hersteller (OEM)-unabhängig sind und einen eigenständigen Markt für Ersatzteile etabliert haben. Basis für die Entwicklung von RAUPE ist daher eine Datenbank von über einer Million gebrauchte Ersatzteile, die als sogenannter digitaler Schrottplatz verwendet wird. Durch data mining Methoden wurden Informationen zu verschiedenen Bauteilen ausgewertet und in RAUPE integriert. Durch RAUPE sollen zum Beispiel für die Werkstatt die Verwendung gebrauchter Ersatzteile verlässlicher gemacht, bzw. durch Garantieansprüche überhaupt erst möglich gemacht werden. Ebenso soll ein Autobesitzer darüber aufgeklärt werden, welchen Umwelteinfluss die Verwendung eines Gebrauchtteils gegenüber eines Neuteils hat.

Acknowledgments. This work was possible through funding the research group Cascade Use by the BMBF with the FKZ 01LN1310A. In addition, we thank callparts recycling for providing the data base on used car parts and Julia Wagner of Kassel University with assisting in some literature studies within her Bachelor thesis in 2017.

5 References

[1] Next Green Car, http://www.nextgreencar.com/ (Accessed 27 Jun 2018)
[2] Umweltbundesamt 2014. Scrap cars, http://www.umweltbundesamt.de/en/topics/waste-resources/product-stewardship-waste-management/scrap-cars#textpart-1 (Accessed 07 Sep 2017)

[3] Bundesministerium für Umwelt, Naturschutz, Bau und Reaktorsicherheit 2017 Legislation in Germany: End-of-Life Vehicle Ordinance, http://www.bmub.bund.de/en/topics/water-waste-soil/waste-management/types-of-waste-waste-flows/end-of-life-vehicles/legislation-in-germany-end-of-life-vehicle-ordinance/ (Accessed 07 Sep 2017)
[4] Altfahrzeugverordnung 2016 Altfahrzeug-Verordnung. Verordnung über die Überlassung, Rücknahme und umweltverträgliche Entsorgung von Altfahrzeugen. Published. on 21.6.2002 (I 2214), https://www.gesetze-im-internet.de/altautov/ (Accessed 07 Sep 2017)
[5] European Commission 2015. Closing the loop - An EU action plan for the Circular Economy. COM(2015) 614 final. European Commission, Brussels, Belgium.
[6] Kalverkamp M, Pehlken A, Wuest T 2017. Cascade Use and the Management of Product Life Cycles. Sustainability, vol. 9 (9), pp. 1540-1563.
[7] Kalverkamp M, Pehlken A 2015. Kaskadennutzung im Automobil – Realität oder Zukunftsmusik? Proc. Konferenz Berliner Recycling und Rohstoffe, http://www.vivis.de/phocadownload/Download/2015_rur/2015_RuR_173-182_Pehlken.pdf (Accessed 06 Oct 2017)
[8] VDI 2014. Innovationspotenzial neuer Werkstoffe für den Automobilbau. Verein Deutscher Ingenieure e.V., Düsseldorf, Germany, https://www.vditz.de/fileadmin/media/news/documents/Studie_Werkstoffinnovationen_fuer_nachhaltige_Mobilitaet_und_Energieversorgung.pdf (07 Oct 2017)
[9] Schmid D, Zur-Lage L, 2014. Perspektiven für das Recycling von Altfahrzeugen – moderne Fahrzeuge und angepasste Recyclingverfahren – Recycling von Altfahrzeugen. Band 7. TK Verlag Karl Thomé-Kozmiensky, Neuruppin, Germany.
[10] Kaerger W, 2014. Markt der Gebrauchtersatzteile. In: Thomé-Kozmiensky K, Goldmann D, 2014. Recycling und Rohstoffe. Band 7. TK Verlag Karl Thomé-Kozmiensky, Neuruppin, Germany.
[11] Eurostat 2014. Reuse and recycling rates in percent of total vehicle weight (W1), http://ec.europa.eu/eurostat/statistics-explained/index.php/File:Reuse_and_recycling_rates_in_percent_of_total_vehicle_weight_(W1),_2014.png (Accessed 09 Oct 2017)
[12] Gruden D, 2008. Umweltschutz in der Automobilindustrie. Motor, Kraftstoffe, Recycling. Vieweg +Teubner / GWV Fachverlage GmbH, Wiesbaden, Germany.

A SWOT and AHP Methodology for the Formulation of Development Strategies for China's Waste EV Battery Recycling Industry

Zhu Lingyun[1] and Chen Ming[1]
[1]School of Mechanical Engineering, Shanghai Jiao Tong University, Shanghai, 200240, P.R. China
mingchen@sjtu.edu.cn

Abstract

Along with the electric vehicle (EV) boom in China, the amount of EV batteries is increasingly growing nowadays, which leads to a dual challenge on waste batteries — the disposal pressure and the recycling demand because of their environmental hazard potentials and recovery value from materials. However, so far, EV battery recycling has not yet been industrialized and a recycling network has not been established in China. In this paper, a SWOT analysis approach is employed to summarize the scrap battery recycling situation in China and sum up the main factors of these four aspects (Strength, Weakness, Opportunity and Threat). Then an Analytic Hierarchy Process (AHP) method is adopted to assess these factors by weight coefficients to identify the key factors in developing a waste EV batteries recycling industry. At last, a TOWS method is used for making development strategies according to the key factors and their weights. The overall strategies include five aspects: laws, economics, system constructions, technology and public education.These strategies, which are presented based on the actual environment of the waste EV battery recycling industry in China and come from qualitative and quantitative methods' analysis, could be useful for China to promote its EV battery recycling management in the future.

1 Introduction

In recent years, the electric vehicle (EV) industry has developed rapidly worldwide, especially in China where people call it new energy vehicle (NEV). As shown in Figure 1, in 2014, which was regarded as the "first year" of EV industry development in China, China became the second largest EV market in the world after the United States [2]. Since 2015, China has been the world's largest electric vehicle market. The proportion of EV sales in China to global sales has increased from less than 10% to around 65%. At the same time, the amount of EV batteries is also increasing. However, the EV battery often has a certain lifetime (about 5-8 years) [6], which will result in a lot of end-of-life EV batteries at some point. According to the prediction of China Automotive Technology and Research Center, the accumulated scrap of EV batteries in China will reach 350,000 tons by 2025 [7]. If not properly disposed of, they will cause environmental pollution and waste of resources. For example, LiFePO4 battery's (LFP) electrolyte decomposition will produce toxic fluoride [8]. On the other hand, Lithium nickel cobalt manganese batteries (NCM) contain the heavy metals Cobalt (Co) and Nickel (Ni), which have recovery value [9].Therefore, the accumulation of scrapped batteries will be a major issue related to both environment and economy. The Chinese government has paid much attention and even attached great importance to it in a new historical height. Recently, China published "New energy vehicle battery recycling Interim Measures" to conduct waste EV battery recycling management [10]. In fact, prior to this regulation, there have been many relative policies to promote battery recycling in China, as shown in Table 3.

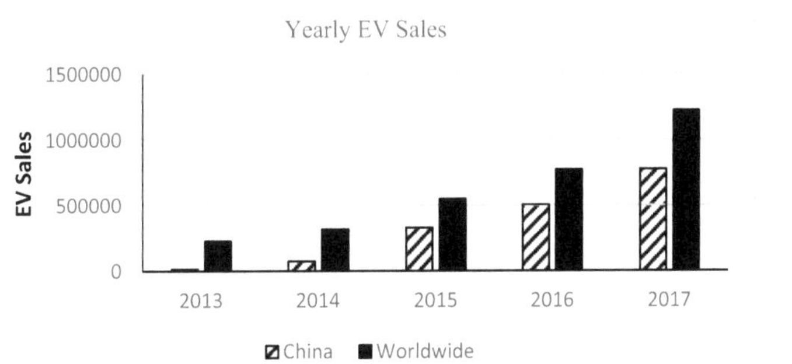

Figure 1. The EV sales between China and Worldwide (year: 2013-2017) [1-5]

However, in today's China, the desire to recycle waste EV batteries has not yet really been carried out, because a recycling network has not yet been established, the recycling industry lacks recycling and disposal specific standards, and the recycling process is not yet mature, let alone that a business model has been formed. On the other hand, regulations and policies related to battery recycling only clarified the responsibilities but did not actually reveal enforcement. A salesperson from a Beijing-based EV 4S store said that car manufacturers did not have a policy on

A. Pehlken et al. (Eds.), *Cascade Use in Technologies 2018*,
https://doi.org/10.1007/978-3-662-57886-5_11

battery recycling. Therefore, the scrapped batteries are usually handled by 4S stores themselves [11]. Consequently, prior to the massive accumulation of scrapped batteries, it is necessary to analyze the recycling situation of China's end-of-life EV batteries and formulate relative development strategies. To the author's best knowledge, not many studies on China's EV battery recycling management strategies exist yet.
In this paper, a SWOT (Strength, Weakness, Opportunity and Threat) analysis approach is employed to summarize the scrap battery recycling situation in China, and an AHP (Analytic Hierarchy Process) method is adopted to assess the keys factors for battery recycling. Then a TOWS method is used to plan the corresponding development strategies. The structure of the paper is organized as follows: Section 2 is an interpretation of the research methods. Section 3 is a SWOT analysis for the situation of the spent EV batteries recycling industry in China. Section 4 is the AHP assessment for the battery recycling development's key factors and the devising of development strategies for them by the TOWS method. Section 5 is the conclusion for this paper.

2 Methodology

This article aims to analyze and evaluate the development prospects of China's waste EV battery recycling and plan a strategic management for the development directions before the recycling industry has reached scale and been industrialized. In consequence, the first step is to conduct a strategic analysis, which is based on the waste EV battery recycling industry's background in China. SWOT analysis is a general method in strategic decision-making, which is derived from the analysis of the internal and external environments of the recycling industry. The internal analysis includes strengths and weaknesses aspects, while the external analysis includes the opportunities and threats aspects [12].The SWOT analysis' results will be represented by a matrix filled with these four aspects and its main factors. In fact, there are some studies on waste management strategies that use SWOT analysis, dealing with MSWM (municipal solid waste management) [13], e-waste [14] and construction waste [15].
However, the SWOT method cannot rank the factors independently by quantitative analysis, which is the basis for strategy-making and prioritizing the development strategies according to the various decision-making factors [16-17]. Analytical Hierarchy Process (AHP) is a suitable method to be used in conjunction with the SWOT matrix [17]. AHP is a classic qualitative and quantitative analysis method for decision-making, which decomposes the relative factors into goal, criteria and alternatives levels [18]. By establishing a hierarchical structure and constructing a comparative judgement matrix, the weight coefficients of each factor are evaluated in the alternatives level. In this paper, the AHP method will rank the factors in the alternatives level to identify the key factors for strategy-making.
In terms of factors ranking, one method is even called A'WOT, which stands for the SWOT and AHP hybrid method. Specifically, this means that SWOT provides the basic frame to perform an analysis of the situation and AHP assists in carrying out SWOT more analytically [19]. Related to end-of-life vehicle recycling research, Junjun Wang and Ming Chen [20] employed the SWOT and AHP method to study the development strategies of used automotive electronic control components recycling industry in China, and Zhang Hongshen and Chen Ming [21], based on this method, conducted research on the recycling industry development model for typical exterior plastic components of end-of-life passenger vehicles. Because of the limited space of the paper and extensive use of AHP method, the principles of AHP are not described in detail.
In regard to strategy-making, by a TOWS analysis method which is derived from the SWOT analysis matrix [22], the development strategies of waste EV battery recycling will be achieved based on the total alternatives weights. The TOWS method [23] is designed to match the environmental threats and opportunities with the company's weaknesses and strengths for development strategy formulation, as Table 1 shows. According to the TOWS method, the strategies consists of four strategic aspects (SO, WO, WT and ST strategies) based on the alternatives level's weight distribution. In detail, SO strategies aim to use its strengths to pursue opportunities, WO strategies intend to take advantage of opportunities to overcome weaknesses, WT strategies are designed to minimize weaknesses and avoid threats, and ST strategies mean to take advantage of the company's strengths to reduce threats.

Table 1. The concept of TOWS method

	Internal Strengths (S)	**Internal Weaknesses (W)**
External Opportunities (O)	**SO:** *"Maxi-Maxi" Strategy* Strategies that use strengths to maximize opportunities	**WO:** *"Mini-Maxi" Strategy* Strategies that minimize weaknesses by taking advantage of opportunities
External Threats (T)	**ST:** *"Maxi-Mini" Strategy* Strategies that use strengths to minimize threats	**WT:** *"Mini-Mini" Strategy* Strategies that minimize weaknesses and avoid threats

In a summary, the research methodology process mainly includes three parts, which are SWOT situation analysis, AHP analysis and TOWS development strategy-making. In the first step, the factors of the waste EV battery recycling situation are integrated into four aspects: Strengths, Weaknesses, Opportunities, and Threats. In the second step, the main factors of these four aspects will be ranked by weight coefficients to identify key factors. In the last step, combination strategies (SO, WO, ST, WT) will be proposed according to the key factors and their weights.

3 SWOT Analysis of Waste EV Battery Recycling in China

In order to conduct a situation analysis, the authors study the external environment and internal environment of waste EV battery recycling in China. According to Hitt's interpretation [24], the former includes the general, industry and competitor environments, and the latter includes resources, capabilities, core competencies, and competitive advantages. Obviously, this classification is based on the perspective of a company's strategic development. However, this paper aims to plan a strategic development at the national level, which will be more inclined to the country's perspective and will also pay attention to the development of the overall EV battery recycling industry. Therefore, similar to the PEST analysis model, the authors integrate environments related to EV battery recycling and divide them into five categories: economics, politics/laws, society, technology, and natural environment. Based on these categories and main factors, the Strengths, Weaknesses, Opportunities and Threats will be summarized.

With regards to the economic aspect, the composition of the EV battery guarantees its recycling value, especially the recovery of Co and Ni metal, which are contained in the electrode as shown in Table 2. However, in China's electric vehicle market, LFP batteries have been widely used in EVs during the past few years, such as SAIC Motor Corp.[1] and BYD Co., Ltd.'s[2] products, which reduce the total recycling value of China's EV batteries. On the other hand, the recovered materials need to undergo some upgrading process before being used as a secondary material in battery production [26], which will cause rising costs for battery recycling. In addition, both the battery manufacturers and the customers worry whether a battery using secondary materials can meet the function and safety requirements [27]. In terms of labor costs, China's abundant labor force is conducive to low costs, but the birth control policy of the past decades leads to a gradual increase in labor costs.

Table 2. Electrode Elements Composition of three types Li-ion Battery [25]

Elements \ Type	$LiNi_{0.8}Co_{0.15}Al_{0.05}O_2$	$LiFePO_4$	$LiMn_2O_4$
Lithium (Li)	1.90%	1.10%	1.40%
Nickel (Ni)	12.10%	0.00%	0.00%
Cobalt (Co)	2.30%	0.00%	0.00%
Aluminum (Al)	0.30%	0.00%	0.00%
Oxygen (O)	8.30%	9.00%	12.40%
Iron (Fe)	0.00%	7.80%	0.00%
Phosphorus (P)	0.00%	4.40%	0.00%
Manganese (Mn)	0.00%	0.00%	10.70%
Titanium (Ti)	0.00%	0.00%	0.00%
Graphite (C)	16.50%	15.30%	16.30%

In the politics/laws aspect, China has comprehensive laws and regulations system related to waste EV battery recycling as Table 3 shows, which includes contents of recycling requirements, responsibility determination, supervision and punishment, etc. The Technical Policy for the Recovery of Automobile Products states that EV manufacturers should be responsible for recycling and treating of sold EV batteries. The Circular Economy Promotion Law of the People's Republic of China regulates the recycling of waste products, such as batteries, which are included in the mandatory recycling list. The Planning for the Development of the Energy-Saving and New Energy Vehicle Industry (2012–2020) regulates the establishment of cascade utilization and recycling management systems for EV batteries. The Automotive Power Battery Industry Specification rules that battery production enterprises must meet the environmental system and occupational health & safety system, recycle or treat the wastes during the manufacturing phase, and cooperate with the vehicle manufacturer for the disposal procedure of used vehicle batteries. The Industry Standard Conditions of New Energy Vehicle Used Battery Utilization and Interim Administrative Measures regulate the relative recycling and disposal technologies and their efficiency requirements, and the distribution of recycling and disposal responsibilities among the stakeholders. The New Energy Vehicle Battery Recycling Interim Measures regulate car manufacturers should be responsible for the recycling and recovery of used batteries and proposes the supervision and management mechanism. However, these legal policies and laws have never been enforced, and there are few detailed

[1] SAIC Motor Corp. is a Chinese vehicle manufacturer in Shanghai

[2] BYD Co. is a Chinese electric vehicle manufacturer in Shenzhen, Guangdong province

implementation rules and standards to follow, although the division of responsibility has already been specified and even the supervision mechanism been proposed.

Table 3. Chinese laws and policies related to waste EV battery recycling

Published year	Legal and policy documents
1989	Environmental protection law of people's republic of China
2001	Hazardous waste pollution prevention technology policy
2003	Waste battery pollution prevention technology policy
2004	Hazardous waste operating license management approach
2006	Renewable resource recovery management regulations
2006	Automotive product recycling technology policy
2008	Circular economy promotion law of the people's republic of China
2012	Planning for the development of the energy-saving and new energy vehicle industry (2012–2020)
2014	Guiding opinions on accelerating promote the use of new energy vehicles
2015	Automotive power battery industry specification
2016	Industry standard conditions of new energy vehicle used battery utilization and interim administrative measures
2016	Specification for comprehensive utilization of waste batteries for new energy vehicles
2016	Electric vehicle battery recycling technology policy
2017	Extended producer responsibility system implementation plan
2018	New energy vehicle battery recycling interim measures

With regards to the technology aspect, the Chinese government has contributed a great deal of support to research on the recycling and cascade use of waste EV batteries, such as National Natural Science Foundation and "863 program". Meanwhile, many environmental protection companies have made significant progress in battery recycling and material recovery, such Brunp Recycling Technology Co.[3], Ltd. and GEM Co., Ltd.[4] . However, the government lags behind the industry's demand for constructing a recycling standards system. Until now, there are only two battery recycling and treatment standards for used EV batteries' residual capacity test and dismantling specification. In addition, the gap in the recycling process between Chinese companies and other international environmental protection companies, such as Umicore, Toxco and OnTo[5], is very obvious. Consequently, China's waste EV battery industrialization is still a long way to go.

Concerning the environment and society aspects, because of the increasing amount of scrapped EV batteries and people's awareness of environmental protection, the battery recycling industry will have a promising development prospect. The government has put much effort into environmental protection infrastructure construction, such as battery recycling bins, but there is no dedicated box for EV batteries recycling. Some environmental protection associations have participated in promoting battery recycling by bringing the relevant stakeholders together to discuss recycling and cooperation mechanisms. For example, the Alliance of Auto Recovery Technology Innovation (AARTI), which is devoted to research and development of general technology for the automobile product recycling industry, including EV and EV batteries in China, was established in 2010 to promote the industry's technological progress in energy conservation, emission reduction and comprehensive utilization of waste resources [21]. Before the industrialization model was eventually formed, some informal small enterprises impacted the battery recycling market by their price advantage, which causes a waste of resources and secondary pollution. Due to the low level of automation in China's battery recycling industry, people's employment problems have been alleviated. But employment in this industry also brings health and secondary pollution problems because the battery's treatment process is not standardized and safe.

Finally, based on the above classification and discussion, a SWOT analysis matrix is formed, as Table 4 shows. Basically, the factors from Opportunities to Threats in the external environment are corresponding to macro policy, environment protection, recycling economy and recycling standards. The factors from Strengths to Weaknesses in the internal environment are symmetrical to recycling market, labor costs, recycling value, recycling technology and waste battery management/cooperation.

[3] Brunp is a Chinese waste vehicle and battery recycling company in Foshan, Guangdong province

[4] GEM is a Chinese waste vehicle and battery recycling company in Shenzhen, Guangdong province

[5] Umicore, Toxco and OnTo are the leading company in resource recovery including waste EV battery recycling

Table 4. The SWOT analysis matrix of China's EV battery recycling industry[6]

Opportunities	Weight	Threats	Weight
O1 Support from national policies and legal systems	0.1874	**T1** Inadequate implementation of policies with little or no penalties	0.0659
O2 Public awareness of environmental protection and waste recycling increased	0.032	**T2** Weak infrastructure and recycling outlets which causes citizens to be unable to participate directly	0.0134
O3 The resources used in the battery are relatively scarce which may have a major impact on the country's economic development	0.0836	**T3** The use of LFP reduces the overall value of recycling which leads to insufficient market forces	0.0263
O4 Recycling standard system is being drafted and formed	0.0320	**T4** Recycling standards are not sound enough to support the entire recycling industrialization	0.0594
Strengths	**Weight**	**Weaknesses**	**Weight**
S1 The number of waste batteries is huge	0.0657	**W1** Recycling network has not yet formed which hinders the development of recycling industrialization	0.0397
S2 Labor costs are relatively low in China	0.0182	**W2** Family planning leads to slowing population growth, a general increase in salary	0.0114
S3 Recycling value of material included in EV battery	0.1129	**W3** Unexplained application prospects of recycled materials	0.0701
S4 Many companies have made breakthroughs in waste EV battery recycling in China	0.0395	**W4** The recycling process has not reached the international advanced level which may lead to recycling value reduced, workers' health threats and secondary pollution	0.1097
S5 Some waste management organizations were established to accelerate internal integration and cooperation in the industry and promote the industrialization of EV battery recycling	0.0138	**W5** Non-qualified recycling companies in the market cause unfair competition which results in unreasonable allocation of resources within the industry	0.019

4 AHP Analysis and Strategies Making for Waste Batteries

4.1 AHP Analysis for Key Factors

In this paper, a hierarchical structure of SWOT factors and AHP analysis is applied as shown in Figure 2. The total environment is the goal level, the internal and external environment and SWOT group form the criteria level, and SWOT factors constitute the alternatives level. After building a clearly structural hierarchy, the next step is to calculate the weight distribution in criteria and alternatives levels. The expert survey method is employed as shown in Table 5, which determines the comparison matrix elements' value. At the criteria level, the internal and external environments are considered 'equally important', as are Strengths and Weaknesses. The Opportunities are regarded as in-between 'equally important' and 'more important'. Therefore, the corresponding weight coefficients are obviously, as shown in Figure 2, according to AHP's 1 to 9 scale ranking of importance. In respect to the alternatives level, four pairwise comparison matrixes (A, B, C, D for S, W, O, T, respectively) should be used to present the results of the expert survey method. Then the maximum eigenvalue (VA, VB, VC, VD) and corresponding eigenvector (TA, TB, TC, TD) of each comparison matrix are calculated as follows.

[6] Because of layout constraints, weights of the SWOT factors, which are calculated and ranked by AHP method in the 4th chapter, are added in table 3.

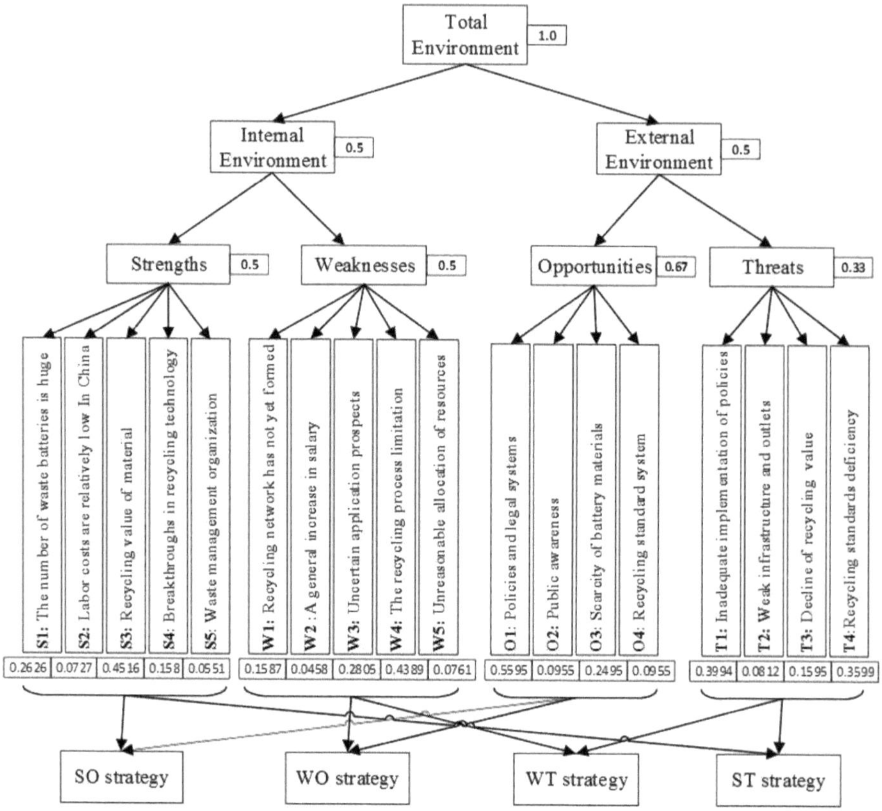

Figure 2. The hierarchical presentation of SWOT and AHP analysis

Table 5. The AHP's expert survey form

Value	1	2	3	4	5	6	7	8	9
Comparison	same importance	median value	more important	median value	obviously more important	median value	strongly more important	median value	extremely more important
Factor X[7]	1	2	3	4	5	6	7	8	9
Factor Y[8]		1/2	1/3	1/4	1/5	1/6	1/7	1/8	1/9

$$A = \begin{pmatrix} 1 & 5 & 1/3 & 2 & 5 \\ 1/5 & 1 & 1/5 & 1/3 & 2 \\ 3 & 5 & 1 & 3 & 5 \\ 1/2 & 3 & 1/3 & 1 & 3 \\ 1/5 & 1/2 & 1/5 & 1/3 & 1 \end{pmatrix} \qquad B = \begin{pmatrix} 1 & 5 & 1/3 & 1/5 & 5 \\ 1/5 & 1 & 1/5 & 1/7 & 1/3 \\ 3 & 5 & 1 & 1/3 & 5 \\ 5 & 7 & 3 & 1 & 7 \\ 1/5 & 3 & 1/5 & 1/7 & 1 \end{pmatrix}$$

$$C = \begin{pmatrix} 1 & 5 & 3 & 5 \\ 1/5 & 1 & 1/3 & 1 \\ 1/3 & 3 & 1 & 3 \\ 1/5 & 1 & 1/3 & 1 \end{pmatrix} \qquad D = \begin{pmatrix} 1 & 5 & 3 & 1 \\ 1/5 & 1 & 1/3 & 1/3 \\ 1/3 & 3 & 1 & 1/3 \\ 1 & 3 & 3 & 1 \end{pmatrix}$$

$$[V_A, V_B, V_C, V_D] = [5.2084, 5.4449, 4.0435, 4.1155]$$

[7] The 3rd row's value is from factor X comparing with factor Y

[8] The 4th row's is from factor X comparing with factor Y

$$[T_A]=\begin{bmatrix}0.4746\\0.1313\\0.8161\\0.2856\\0.0996\end{bmatrix},\ [T_B]=\begin{bmatrix}0.262\\0.0664\\0.4321\\0.8541\\0.1035\end{bmatrix},\ [T_C]=\begin{bmatrix}0.8919\\0.1522\\0.3977\\0.1522\end{bmatrix},\ [T_D]=\begin{bmatrix}0.7048\\0.1433\\0.2814\\0.6352\end{bmatrix}$$

Prior to the weight coefficients calculation, the consistency of the comparison matrix needs to be examined based on the formula (1) and (2). If the C.R. value is less than 0.1, the comparison matrix will be a satisfactory consistency matrix, and the corresponding eigenvector will be used as the initial weight coefficient after being normalized. If not, the comparison matrix will be reviewed and revised.

$$C.I. = (V - n)/(n-1) \quad (1)$$
$$C.R. = C.I./R.I. \quad (2)$$

Where,
V is the maximum eigenvalue.
n is the order of the comparison matrix.
C.I. is the consistence index.
R.I. is the random consistency index whose value is shown in Table 5.
C.R. is the consistence ratio.

Table 6. The value of random consistency index [18]

n	3	4	5	6	7	8	9	10	11
R.I.	0.58	0.89	1.12	1.24	1.32	1.41	1.45	1.49	1.52

According to formula (1) and (2), the consistency test's results showed that the C.R. value are 0.0465, 0.0993, 0.0163 and 0.0433 for Matrix A, B, C and D, respectively, which are all under 0.1. But we can see that the C.R. value of matrix B is too close to 0.1, so matrix B should be revised slightly. Specifically, after rethinking and deciding, the authors enhance the importance of recycling network (W1) to recycling technology (W4) from 1/5 to 1/3, reduce the importance of recycling network (W1) to allocation of resources (W5) from 5 to 3, raise the importance of labor costs (W2) to recycling technology (W4) from 1/7 to 1/5, and decrease the importance of recycling technology (W4) to allocation of resources (W5) from 7 to 5. Then the maximum eigenvalue ($V_{B'}$) and corresponding eigenvector ($T_{B'}$) are shown below, and the C.R. value is 0.0873 which is more reasonable for a satisfactory consistency matrix.

$$B' = \begin{pmatrix}1 & 5 & 1/3 & 1/3 & 3\\1/5 & 1 & 1/5 & 1/5 & 1/3\\3 & 5 & 1 & 1/3 & 5\\3 & 5 & 3 & 1 & 5\\1/3 & 3 & 1/5 & 1/5 & 1\end{pmatrix}\ V_{B'} = 5.3909,\ [T_{B'}]=\begin{bmatrix}0.2877\\0.083\\0.5085\\0.7955\\0.1379\end{bmatrix}$$

Finally, the initial weight coefficients can be calculated by standardization of eigenvectors (T_A, $T_{B'}$, T_C, T_D) and the results are shown in Figure 2. Meanwhile, the total weight coefficients of the alternatives level can be computed based on all levels' initial weight rank by multiplication. Results are shown in Table 4.
Generally speaking, the quantitative analysis in the AHP method is derived from the experts' experienced scoring which is more or less subjective. However, it is efficient and relatively reasonable for decision-making as waste battery recycling has not yet been industrialized and a recycling network has not yet been established in China. Along with the changes and development of the industrialization of EV battery recycling, the main factors in SWOT analysis may be updated and replaced, and the development strategies should adjust accordingly.

4.2 TOWS Analysis for Strategies-Making

As the results of the total weight SHOW, seven factors' coefficient is above 0.65, one factor is close to 0.06 and THE others are below 0.04. In order to formulate more effective and direct development strategies for waste EV battery recycling, the lower weight factors are ignored and the higher weight factors are used for devising strategies as shown in Table 4. These are large amount of waste EV battery (S1, 0.0657), recycling value of material (S3,0.1129), unexplained application prospects (W3,0.0701), recycling process limitation (W4,0.1097), policies and legal systems (O1,0.1874), scarcity of battery materials (O3,0.0836), inadequate implementation of policies (T1,0.0659), recycling standards deficiency (T4, 0.0594).
Finally, the TOWS analysis is employed, as Table 1 shows, to generate four combined strategies (SO, ST, WO and WT) based on the matching and analyzing of these eight factors with higher weights.

SO strategies: utilize support from laws and policies to promote waste battery collection and recycling.

- Formulate more detailed plans for waste EV battery recycling and enforce the implementation for extended producer responsibility (EPR).
- Encourage recycling behavior of producers and their partners through subsidies or tax reductions.
- The government invests in more infrastructure to serve the recycling of EV battery.
- The government shuts down non-qualified and non-standard battery recycling companies to make resources flow to qualified companies.
- The government funds scientific research projects for companies, universities and research institutes, and accelerates the transformation of research results to improve recycling technology and value.
- The government instills the idea of green production and circular economy into the companies, and encourages the construction of strategic organizations or alliances for the battery recycling technology development.

ST strategies: continue to deepen the recycling implementation policies and improve processing standards to promote battery recycling.

- The government establishes a recycling system based on EPR principles and strengthens the construction of an information traceability system to improve the definition of responsibility and the level of reverse logistics management.
- The government promotes awareness of environmental protection to citizens and launches the whole society to supervise.
- The relevant government departments continue formulating recycling processing standards to support the development of the battery recycling industry.
- The government establishes more detailed treatment standards on all steps of reverse logistics to support the battery recycling industrial chain.

WO strategies: Strengthen the technical cooperation of recycling process and look for the downstream market of secondary materials.

- The government encourages the reuse of secondary materials and promulgates related management measures and enterprise access conditions.
- The government assists battery recycling companies to explore the recovery market for recycled materials.
- Develop cascade use of waste batteries to other areas to increase their overall value.
- Recycling companies strengthen their cooperation with domestic and foreign battery recycling corporations to introduce the advanced experience in the recycling process and recycling value.

WT strategies: Strengthen the application of green design theory and increase safety awareness.

- Through the recovery rate and emission requirements, manufacturers are encouraged to implement the application of green design theory, such as recyclability design and detachability design to promote the recycling of spent EV batteries.
- For the unclear application prospects of recycled materials, the recycled coarse materials (raw materials and precursors) can be resold to foreign companies for initial recovery value. Meanwhile, the companies strengthen technology exchanges and cooperation to gradually master higher level recycling values.
- Enterprises improve the standardization of waste EV battery management process, and train workers' safety awareness.

To sum up, the development strategies of waste EV battery recycling can be mainly divided into five aspects including laws, economics, system constructions, technology and public education as shown in Figure 3.

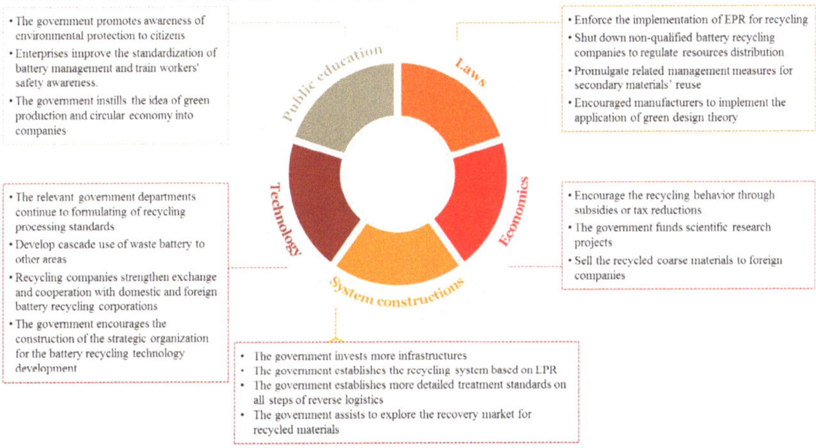

Figure 3. The total development strategies of waste EV battery recycling industry in China

5 Conclusion

In the wake of the development of electric vehicles in China, the problem of EV battery scrapping has become increasingly serious. Based on the internal and external environment of the waste EV battery recycling industry in China, this paper conducts a SWOT situation analysis and summarizes the main factors of Strengths, Weaknesses, Opportunities and Threats. In addition, an AHP method is employed to rank these factors by the weight coefficients, and based on this, a TOWS method is adopted to make development strategies for waste EV battery recycling in China. These can be summarized by five aspects: laws, economics, system constructions, technology and public education. Because the recycling of waste EV batteries has not yet been industrialized and the recycling network has not yet been established in China, it is necessary to propose development strategies for decision makers.

Along with the changes and development of the industrialization of EV battery recycling, the main factors in SWOT analysis may be updated and replaced, and the development strategies should adjust accordingly. In general, this paper combines the SWOT method for situation analysis, the AHP method for factors ranking, and TOWS method for strategy-making, which is a systematic and flexible approach structure. The research process in this article can be further extended to WEEE, waste EV and waste portable battery recycling fields.

6 Zusammenfassung

Mit dem Boom der Elektromobilität nimmt in China auch die Zahl der Antriebsbatterien zu, was die Behandlung der ausgedienten Batterien aufgrund des Entsorgungsdrucks durch potenzielle Umweltgefahren und die Recyclingforderungen zur Wiedergewinnung wertvoller Materialien zu einer zweifachen Herausforderung werden lässt. Bisher findet in China jedoch kein Batterierecycling auf industriellem Level statt und es konnte kein Recyclingnetzwerk etabliert werden. In diesem Beitrag wird eine SWOT Analyse durchgeführt, um die aktuelle Situation des Batterierecylings in China zusammenzufassen und die Hauptfaktoren der vier Aspekte (Stärken, Schwächen, Chancen und Bedrohungen) aufzuzeigen. Zur Gewichtung und Identifizierung der Schlüsselfaktoren in der Entwicklung einer Batterierecyclingindustrie wird anschließend ein analytischer Hierarchieprozess (AHP) durchgeführt. Basierend auf diesen Schlüsselfaktoren werden abschließend mittels der TOWS Methode Strategien entwickelt, die die fünf Aspekte Gesetze, Wirtschaft, Systemkonstruktion, Technologie und öffentliche Aufklärung beinhalten. Diese auf der tatsächlichen Situation des Batterierecycling basierenden sowie mittels quantitativer und qualitativer Methoden entwickelten Strategien können zur Förderung eines Batterierecyclingsystems in China genutzt werden.

Acknowledgements. The authors express their sincerest thanks to the National Natural Science Foundation of China for financing this research within the program" Fundamental Research on Catalytic Gasification of Automobile Shredder Residues (ASR): Mechanism and Its Recovery" under the Grant No. 51675343.

7 References

[1] China Association of Automobile Manufacturers. http://www.caam.org.cn/
[2] ResearchInChina. http://www.researchinchina.com/
[3] New Energy Automotive News Ev. http://www.sohu.com/a/58091045_372664
[4] D1EV. https://www.d1ev.com/news/shuju/48831
[5] D1EV Research Institute. Global New Energy Vehicle Industry Development Report (2017). Technical report, D1EV

[6] How long is the EV battery's life? http://www.sohu.com/a/76155495_377286
[7] D1EV. https://www.d1ev.com/kol/54717
[8] Niedzicki, L., Kasprzyk, M., Zukowska, G. et al. (2009). New conductive salts as potential lithium battery electrolytes tested in pc and gel-pc system. Gastroenterology, 144(5), S-314-S-314.
[9] Lupi, C., Pasquali, M., & Dell'Era, A. (2005). Nickel and cobalt recycling from lithium-ion batteries by electrochemical processes. Waste Management, 25(2), 215-220. doi:10.1016/j. wasman.2004.12.012
[10] Ministry of Industry and Information Technology of the People's Republic of China. http://www.gov.cn/xinwen/2018-02/26/content_5268875.htm
[11] National Business Daily. http://auto.163.com/17/0707/08/CONRI4D8000884N3.html
[12] Srivastava, P. K., Kulshreshtha, K., Mohanty, C. S.et al. (2005). Stakeholder-based SWOT analysis for successful municipal solid waste management in Lucknow, India. Waste Management, 25(5), 531-7. doi:10.1016/j.wasman.2004.08.010
[13] Zotos, G., Karagiannidis, A., Zampetoglou, S. et al. (2009). Developing a holistic strategy for integrated waste management within municipal planning: challenges, policies, solutions and perspectives for hellenic municipalities in the zero-waste, low-cost direction. Waste Management, 29(5), 1686-1692. doi:10.1016/j.wasman.2008.11.016
[14] Pariatamby, A., & Victor, D. (2013). Policy trends of e-waste management in Asia. Journal of Material Cycles & Waste Management, 15(4), 411-419. doi: 10.1007/s10163-013-0136-7
[15] Yuan, H. (2013). A swot analysis of successful construction waste management. Journal of Cleaner Production, 39(5), 1-8. doi: 10.1016/j.jclepro.2012.08.016
[16] Shojaei, M., Abbaszade, S., & Somayeh Aghaei, S. (2013). Using analytical network process (ANP) method to prioritize strategies resulted from SWOT matrix case study: Neda Samak Ashena Company. Interdisciplinary Journal of Contemporary Research in Business, 4(9), p603
[17] Shahabi, R. S., Basiri, M. H., Kahag, M. R., & Zonouzi, S. A. (2014). An ANP–SWOT approach for interdependency analysis and prioritizing the iran's steel scrap industry strategies. Resources Policy, 42(42), 18-26.doi: 10.1016/j.resourpol.2014.07.001
[18] Saaty, T. L. (2001). Analytic hierarchy process. 109-121.doi: 10.1007/1-4020-0611-X_31
[19] Kangas, J., Pesonen, M., Kurttila, M., & Kajanus, M. (2001). A' WOT : Integrating the AHP with SWOT Analysis. Isahp, 189–198. doi:10.1007/978-0-387-76813-7
[20] Wang, J., & Chen, M. (2012). Management status of end-of-life vehicles and development strategies of used automotive electronic control components recycling industry in china. Waste Manag Res, 30(11), 1198-207.doi: 10.1177/0734242X12453976
[21] Zhang, H., & Chen, M. (2013). Research on the recycling industry development model for typical exterior plastic components of end-of-life passenger vehicle based on the swot method. Waste Manag, 33(11), 2341-2353. doi:10.1016/j.wasman.2013.07.004
[22] Weihrich, & Heinz. (2010). Management: a global and entrepreneurial perspective/13th ed. Economic Science Press.
[23] Weihrich, H. (1982). The tows matrix—a tool for situational analysis. Long Range Planning, 15(2), 54-66. doi:10.1016/0024-6301(82)90120-0
[24] Hitt, M. A., Ireland, R. D., & Hoskisson, R. E. (2013). Strategic management: competitiveness & globalization. South-Western Cengage Learning
[25] Gaines, L., Sullivan, J., & Burnham, A.(2011). Life-cycle analysis for lithium-ion battery production and recycling. Argonne National Laboratory
[26] Dunn, J. B., Gaines, L., Kelly, J. C., & Gallagher, K. G. (2010). Life cycle analysis summary for automotive lithium-ion battery production and recycling. doi:10.1007/978-3-319-48768-7_11
[27] Amarakoon, S., Smith, J., & Segal, B. (2013). Application of life-cycle assessment to Nano scale technology: lithium-ion batteries for electric vehicles. Evaluation. No. EPA 744-R-12-001.

Evaluation of the Recyclability of Traction Batteries Using the Concept of Information Theory Entropy

Nicolas Bognar[1,2], Julian Rickert[1,2], Mark Mennenga[1,2], Felipe Cerdas[1,2], Christoph Herrmann[1,2]
[1] Chair of Sustainable Manufacturing & Life Cycle Engineering, Institute of Machine Tools and Production Technology (IWF), Technische Universität Braunschweig, Braunschweig, 38106, Germany
[2] Battery LabFactory Braunschweig (BLB), Technische Universität Braunschweig, Braunschweig, 38106, Germany
n.bognar@tu-braunschweig.de

Abstract

As traction battery technologies and electro mobility as a whole continue to grow in importance, the recyclability of batteries has increasingly gained attention in politics, industry and science. The aim of this paper is to broaden the understanding about the recycling of traction batteries by applying the concept of information theory entropy. To this end, information theory-based entropy indicators are used to determine the material mixing complexity of current and future battery chemistries used in electric vehicles. Through the integration of different economic metrics and with the help of additional related information on industrial, political and social influencing factors the recyclability of traction batteries is evaluated and the development of future battery recycling systems and policies is discussed. The results show that the proposed methodology is suitable for comparing different product technologies and that significant differences exist regarding the determining factors for the recyclability of different battery technologies.

1 Introduction

The electrification of transportation is increasingly seen as a solution towards more sustainable mobility. While traction batteries play a fundamental role within the field of electro mobility, their manufacturing and the production of the materials required pose various economic challenges [1, 2]. As of now, traction batteries are made of materials like Lithium, Cobalt and Nickel, which are associated with relatively high material costs and material criticality due to economic importance and geographic concentration [3]. Thus, the secondary raw material streams from the recycling of electric vehicle (EV) batteries may become an important source of materials for new traction batteries. If consistently implemented and executed, recycling might contribute to reducing the demand for primary raw materials [4].

The aim of this paper is to broaden the understanding about traction battery recycling, to discuss a method for determining the recyclability and to identify potential influencing factors in the development of future battery recycling systems and policies at an early development stage. The presented approach builds on existing research by Dahmus and Gutowksi [5] regarding the recyclability of products and applies their proposed methodology for traction batteries. A structured framework to gain relevant information for the recycling of traction batteries is introduced and the recyclability of different battery cell chemistries is compared systematically.

2 Battery Recycling

The transformation of the mobility sector towards electric drivetrains powered by batteries has started only recently within the last decade [6]. Therefore, a significant stream of spent battery packs is estimated to become available within the next ten years. As of today, most Li-Ion batteries are used for consumer electronics products with significantly smaller battery sizes than EV batteries. Recycling technology and capacity exists for batteries of other applications, but is relatively new for EV batteries. Currently, there are only few dedicated industrial recycling facilities. However, more capacities are planned due to the predicted increasing demand over the coming years.

Battery recycling generally uses mechanical, pyrometallurgical and hydrometallurgical processes, usually used in combination [7]. As a first step, there are pre-treatments of battery packs and modules, such as deep discharge and disassembly of the peripheral parts like the housing, cables and power electronics. This step is followed by mechanical processes (shredding, sorting, drying, sieving, etc.) with subsequent pyrometallurgical and hydrometallurgical treatments. Melting processes have already been successfully implemented on an industrial scale, as they are the most economical solution for the currently small stream of spent batteries. They are also the most robust processes for the heterogeneous input waste streams. There is a variety of different technologies, sizes and geometries for current and future EV batteries [7]. This growing product diversity, complexity and resulting material dilution within batteries is threatening to become a major obstacle in successful industrial battery recycling. These factors are equally relevant for the upstream disassembly of battery packs and the following treatment of the active material of battery cells.

To ease disassembly and recycling of complex products, well established methodologies like Design for Recycling (DfR) and especially Design for Disassembly (DfD) strategies and guidelines [8] are available, e.g. by standardizing module geometries or joining technologies and locations. This way the valuable battery module and pack housing materials like steel or Aluminum can be recovered and the battery cells separated. However, a viable option to improve the battery cell

A. Pehlken et al. (Eds.), *Cascade Use in Technologies 2018*,
https://doi.org/10.1007/978-3-662-57886-5_12

material recycling processes could be to specialize the processes on those material compositions which have the best recyclability. To determine and compare the recyclability of products, so called entropy indicators have been applied successfully, e.g. in the case of electronics [9] or photovoltaic modules [10]. This paper provides a case study on the recyclability of EV battery systems with the help of material complexity indicators based on the Shannon entropy from information theory [5]. Consequently, the underlying methodology is presented in the following chapter.

3 Methodology

With regards to the aim of this paper and under consideration of existing approaches this paper provides a methodology for determining the recyclability of current and future battery technologies (see Figure 1). It builds on prior research by Rechberger and Brunner [11] as well as Gutowski and Dahmus [5, 12], who use information theory entropy to describe the recyclability of products. It consists of six steps with a cascading character, as every information output is used in the subsequent step. The 1st step of the methodology serves the identification of relevant technologies based on the evaluation of technology roadmaps, in order to determine the technological scope of the analysis. Based on the technological scope, a material analysis is conducted in the 2nd step. Material inventories and stoichiometric calculation of mass fractions are used to provide product compositions and material concentrations. This information is included in the 3rd step together with material prices and the Sherwood methodology to identify those materials which are targeted in recycling. Thereafter, the Shannon entropy is calculated together with the sum of single recycled material values. The 5th step integrates further economic metrics to derive a comprehensive economic perspective. With the help of additional related information on industrial, political and social influencing factors, the 6th and last step serves the overall analysis and discussion of strategies on how to increase the recyclability of the relevant product. In the following, the six steps of the methodology are presented in detail.

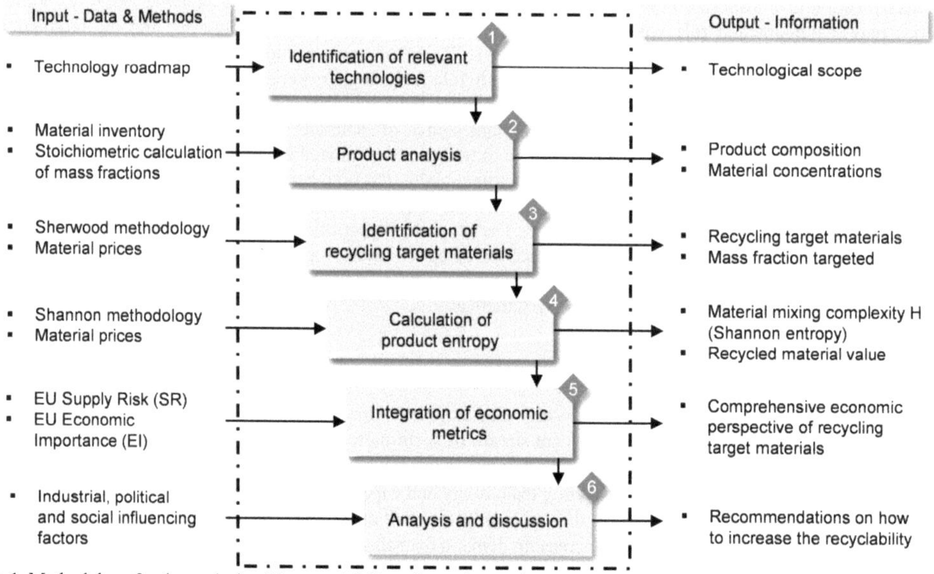

Figure 1. Methodology for the analysis of the recyclability of complex products

Step 1: Identification of Relevant Technologies

The 1st step is a technology screening and identifies the relevant technologies from a selection of market studies, technology roadmaps and expert knowledge. Market studies and experts provide valuable information about the technologies that are relevant to the industry and for recycling. The technology roadmaps provide information about the development of the technology within the next years. Future developments should be considered in the analysis when the product life cycle is expected to last relatively long, since it results in a temporal shift for the waste streams to enter recycling. This is especially the case for the battery cell technology development. The output of this step is the technological scope and should contain all relevant product technology variants.

Step 2: Product Analysis

The 2nd step is the analysis of the components and mass fractions of the identified technologies. Preferably, real inventory data should be used for the material analysis in order to get realistic results. Experience shows that models tend to provide an optimized material inventory that includes less peripheral parts than real recycling studies, especially for battery technology. This is mainly due to over-engineering of safety housing parts for increased product safety. However, in some cases models need to be used, because of lacking published data or data for future technologies that does not exist at the time of performing the study. Based on mass distributions of components and their material compositions, mass fractions of each material can be calculated on component and product level. Stoichiometric calculations are an important tool for this.

Step 3: Identification of Recycling Target Materials

The 3rd step uses the results of the second step, detailed product compositions and exact material concentrations, as well as corresponding material prices and the Sherwood methodology [13] in order to identify materials that should be targeted during recycling, as proposed by Dahmus and Gutowksi [5]. Sherwood [13] indicated that the selling prices of virgin materials vary approximately proportionally with their degree of concentration in the matrix from which they are extracted. Figure 2 on the left shows the relationship between the concentration of a target material in the extraction matrix and the market value of a target material for metals and medicine products. The underlying idea using the Sherwood plot is that materials, which lie above the Sherwood line, are potential candidates for recycling; materials that lie beneath the Sherwood line are considered not valuable enough or too diluted for extraction and recovery. Allen and Bemanesh [14] transferred this approach to examine the economic potential of industrial waste streams and Johnson et al. [9] proved that the Sherwood plot is useful for predicting which materials to target when electronics like mobile phones and personal computers get recycled. This can be seen in Figure 3 on the right, as historically targeted materials lie above the Sherwood line.

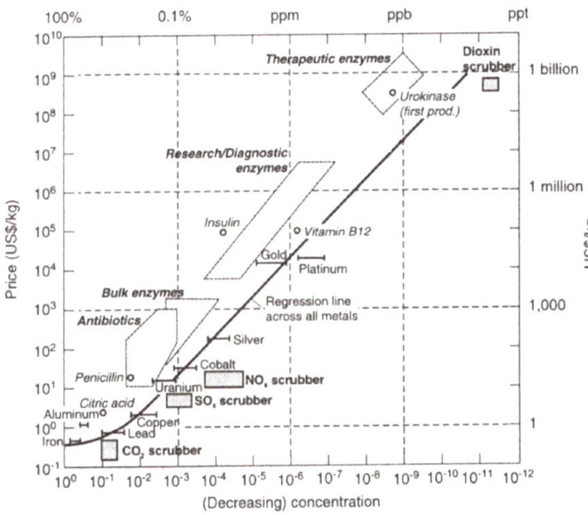

Figure 2. Sherwood plot showing the relationship between the concentration of a target material in a feed stream and the market value of the target material [15]

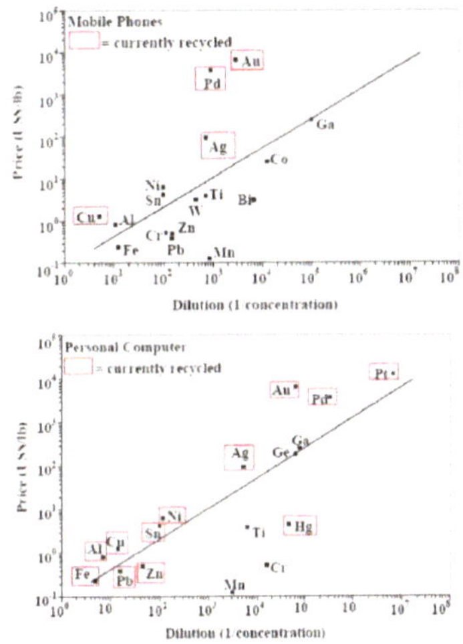

Figure 3. Use of Sherwood plots in the context of electronics recycling [9]

Step 4: Calculation of Product Information Theory Entropy

In the 4th step, the previous results – the information about which materials to target – as well as the material prices are used within Shannon's noiseless coding theorem, which is a method originally used in information theory and is based on L. Boltzmann's statistical description of entropy. In information theory, it is used to measure the loss or gain of information about a system, whereas in statistics it is used to measure the variance of a probability distribution. In the context of material recycling, the Shannon entropy is used as a proxy for material mixing complexity in order to assess the separation and recovery efforts of pure substances from the product. It was first adapted to material recycling by Dahmus and Gutowski [12] and later on expanded by Mohamed Sultan, Lou and Mativenga [16]. A related indicator, the Rényi entropy, was used by Fthenakis and Anctil [10] in a study on photovoltaic panels. Here, the Shannon entropy H is a function of the number of component materials in a mixture M and the mass fraction of each material of the total composition c_i. Therefore, H can be calculated using Formula 1:

$$H = -\sum_{i=1}^{M} c_i \log_2 c_i \text{ [bits]} \tag{1}$$

Gutowski and Dahmus [12] plotted the relationship between the sum of the recycled material's value and the material mixing complexity H and introduced an 'apparent recycling boundary' as shown in Figure 4 [5].

The output of the 4th step can be seen in Figure 4. The expansion by Mohamed Sultan and colleagues [16] was not deemed applicable for traction battery technologies. The authors integrate the recycling Technology Readiness Level (TRL) and Material Security Index (MSI) into their methodology. For some cases the expansion will not bring any additional information, as the TRL of relevant recycling technologies is similar and because the MSI aggregates a great number of Key Performance Indicators (KPI) from different fields into a single numerical value and can therefore produce misleading results. This is the case for the battery technology analysis. Nevertheless, further factors were included into the analysis, as material mixing complexity provides a good indication of product recyclability, but is not the only determining factor. There are manifold external influences onto the recyclability. These influencing factors include the availability of recycling technology, availability of recycling waste streams, efficiency of recycling technology, demand and supply for materials, political situation, product material criticality, etc. The majority of those factors can be categorized as economic influences.

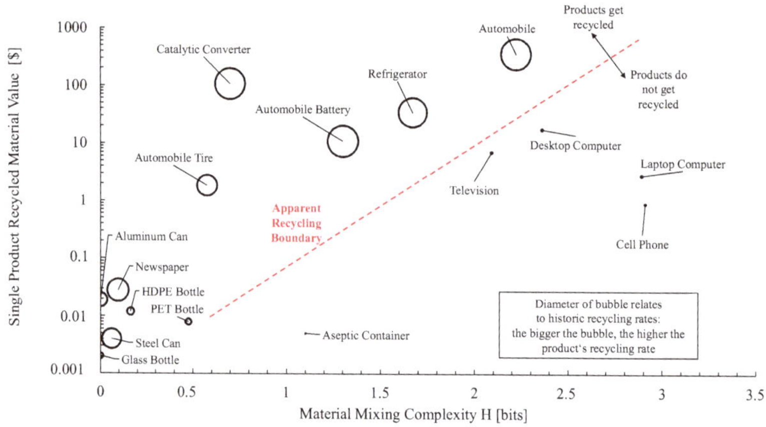

Figure 4. Relationship between recycled material value, material mixing complexity, recycling rate (data from 2007) and apparent recycling boundary [12]

Step 5: Integration of Economic Metrics

The 5th step includes economic metrics into the analysis. The economic criticality of a material can be expressed through different indicators. The EU uses two metrics to assess the criticality of a material, the Economic Importance (EI) and the Supply Risk (SR) [17]. The EI is a quantifiable metric of the relevance of a specified material for the European industry. It takes into account the share of consumption of a material in end use applications and the value added of the corresponding manufacturing sectors. It is adjusted by the substitution indicator that addresses the interchangeability of materials with similar costs and technical properties. The indicator is scaled to a range from 0 to 10, with a higher value indicating a higher economic importance for the European economy [17]. The SR on the other hand addresses the risk of a material supply disruption in the EU and consists of three factors. It considers the political stability and level of concentration of the producing respectively material processing countries. Furthermore, it addresses the substitutability of a material for the economy based on a weighted average calculation over all industries. Additionally it includes the extent to which the European demand for a material is supplied from recycling [17]. These metrics are adopted in this study, since on the one hand they provide insights into the manufacturing sector, which is assumed to grow over the next years for battery manufacturing. One the other hand, the SR incorporates the material scarcity, political stability of producing countries, existing material recycling capabilities and how well these materials can be substituted by similar materials. These factors are highly relevant for the materials for battery cells, specifically Cobalt, Nickel and Lithium [18].

Step 6: Analysis and Discussion

Finally, the 6th step analyses and discusses the results of the previous steps based on a comprehensive perspective. Additional information such as market trends, technology development, material price predictions, expected regulations or possible business models is put into the context of the analysis. The influence of these factors onto product recyclability is discussed and summarized. The results of the analysis are projected onto guidelines or strategies to increase the recyclability of the product technology. Restricting as well as supporting factors are identified from an economic and political perspective.

4 Case study: Batteries for Electric Vehicles

4.1 Step 1: Identification of Relevant Technologies

Traction batteries used currently within electric vehicles are based on Lithium-Ion (Li-Ion) technology. There are different chemistries used for electric and plug-in hybrid electric vehicles (PHEV) in different countries and for different EV brands. Generally, these chemistries are used as cathode material coated onto an Aluminum current collector combined with a Graphite coated Copper anode. The main technologies currently used in EV are NMC, NCA, LFP and LMO respectively LMO/NMC. NCA ($LiNi_{0.8}Co_{0.15}Al_{0.05}O_2$) is used mostly by Tesla. NCA has the advantage of providing a high capacity and voltage, but is more vulnerable to safety problems and is expensive [19]. LFP ($LiFePO_4$) based cell chemistries are mostly used by Chinese EV manufacturers [20]. They provide a high cycle life but low energy densities. Very high energy densities within Li-Ion batteries are currently achieved by NMC (NMC111 – $LiNi_{0.33}Mn_{0.33}Co_{0.33}O_2$) based cell chemistries, used in most EV [6]. However, NMC cells are relatively expensive due to their high shares of Nickel and Cobalt [2]. The NCM technology is evolving at a fast pace. For this study, a material inventory was considered based on a conventional EV battery provided by Diekmann et al. [3]. Cerdas et al. [21] provide a more recent material inventory for an energy optimized NMC cell, which is characterized by a higher specific energy or energy density. In order to identify both inventories in the figures, the inventory provided by Cerdas et al. is named high energy (HE)-NMC. Research for this chemistry aims at increasing the share of Nickel in the cathode towards currently already used NMC622 ($LiNi_{0.6}Mn_{0.2}Co_{0.2}O_2$) and eventually NMC811 ($LiNi_{0.8}Mn_{0.1}Co_{0.1}O_2$) and to add silicon to the anode of the battery [22]. In the past, LMO ($LiMn_2O_4$) has been used for EV. Due to the better performance of the state-of-the art technologies, it is now less common to be found within EV batteries [6].

The battery industry is characterized by fast technological developments. New chemistries are developed within short time frames with expected market introductions in the coming years until 2030. One example for next generation batteries is the Lithium Sulphur (Li-S – LiS_8) battery, which may see use within EV once the low cycle life can be improved and the challenge of the high volume can be overcome [21]. Additionally, post-Lithium batteries, such as Sodium Ion (Na-Ion – $Na_{1.1}Ni_{0.3}Mn_{0.5}Mg_{0.05}Ti_{0.05}O_2$) batteries, may see their market introduction in the next decade [23]. Next to batteries for EV, there is a variety of chemistries used for other applications, e.g. stationary applications. These are not taken into consideration for this study. Table 1 provides an overview of the technological scope of this study, respectively the considered battery types and their key attributes.

Table 1. Technological scope of the study – Considered battery types

Battery Type	Mass [kg]	Specific energy [Wh/kg]	Format	Data Source Type	Source
NMC	350	105	Pouch	Real battery	[3]
HE*-NMC	340	150	Pouch	Real battery	[21]
LMO/Gr	63	130	PHEV (prismatic)	Real battery	[19]
LMO/Ti	106	130	PHEV (prismatic)	Real battery	[19]
NCA/Gr	76	110	PHEV (prismatic)	Real battery	[19]
LFP/Gr	82	90	PHEV (prismatic)	Real battery	[19]
Na-Ion	343	102	18650 (cylindrical)	Battery model	[24]
Li-S	340	150	Pouch	Battery model	[21]

*energy optimized NMC cell (HE: **H**igh **E**nergy)

4.2 Step 2: Product Analysis

Spent batteries usually enter the recycling waste stream as battery packs. Additionally to the cells, these battery packs consist of a battery management system (BMS), a cooling system, power electronics and sensors as well as structural elements for mechanic stability and safety [25]. These components are part of the module and pack periphery that consists mainly of steel, Aluminum and plastics, which in total can make up to 25-45% of the weight of NMC battery packs [3].

Cells consist of the components anode, cathode, separator, electrolyte and cell housing. Each cell technology uses different materials for its components. The anode for most cells consists of a Copper current collector foil, which is coated with slurry consisting of Graphite, binder, Carbon Black and a solvent that fully evaporates during the production process. Technologies with different anodes are Li-S with a Lithium metal anode and LMO/Ti with spinel structured Li-Titanate anode. The key differences between the cells are on the cathode side. All cells use an Aluminum current collector foil, which is coated by different cathode slurries. Similar to the anode, the cathode slurry consists of the active material, binder, Carbon Black and solvent. For each cell chemistry, the material composition of the cathode material is calculated using stoichiometric calculations. The separators used in battery cells are based on either polymer or ceramic. There is a variety of electrolytes available and used within cells. However, they cannot be recovered from recycling with the currently available recycling technologies. The cell housing typically consists of Aluminum or steel for hard case cells and a composite Aluminum-polymer foil for pouch cells. A typical distribution of materials is shown in Figure 5 for a NMC battery system.

Based on the results for the material analysis for each cell, the material inventories are set up. They contain the cell materials and their concentration within the cell. The results are shown in the Sherwood plots in Figure 6, with the dilution as the inverse of the concentration computed on the x-axis. The most common materials within most cells are Copper, Aluminum and Graphite. According to the cell chemistry, other metals such as Nickel, Cobalt, Manganese and Titanium can be identified. Generally, the Lithium content that can be found within cells is relatively low compared to other elements.

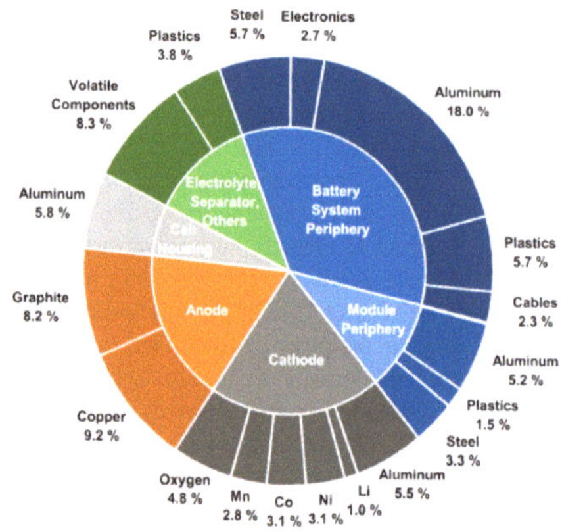

Figure 5. Generic composition of a NMC EV battery system [3]

4.3 Step 3: Identification of Recycling Target Materials

As the level of details for the provided material inventory is different for each study, it is necessary to first define a consistent material counting scheme. This is derived from the Sherwood plots in Figure 6. Materials above or relatively close to the Sherwood line are considered sufficiently rewarding for targeting during recycling. However, other arguments are also considered for the selection of materials, e.g. the effort to recover the materials from the recycling stream. The red underlines in Figure 6 mark the materials that are targeted during recycling. These materials and their concentration are considered in the calculation of the material mixing complexity. The Sherwood plots in Figure 6 show that Lithium, due to its low concentration, may not be within the material with the highest recycling priority from a cost and dilution perspective. Aluminum, Cobalt, Nickel, Manganese and Copper generally have concentrations high enough to be considered as target materials for recycling. Sodium and Sulphur from the new battery generations have a relatively low market value and therefore are not considered for recycling. Since the electrolyte cannot be recovered with a sufficient quality in current recycling processes [3], the values have been neglected. The ratio between prices of virgin and recycled materials was established based on average price differences between new and scrap materials by Anctil and Fthenakis [10], who found that on average a recycled material is worth 60% of a virgin material. This assumption has been adopted for the present case study. The underlying data for the analysis is shown in Table 2.

Table 2. Material data and sources

Material	**Price** (virgin) [€/kg]	**Price** (recycled) [€/kg]	**Economic Importance** (EI)	**Supply Risk** (SR)	**Source** (Price)	**Source** (EI & SR)
Al	3.51	2.11	6.5	0.5	[26]	[27]
Co	33.61	20.16	5.7	1.6	[26]	[28]
Cu	7.87	4.72	4.7	0.2	[26]	[27]
Fe	0.61	0.37	6.2	0.7	[26]	[27]
C	1.29	0.77	2.9	2.9	[29]	[28]
Li	5.56	3.34	2.4	1	[30]	[27]
Mg	4.26	2.55	3.7	0.7	[26]	[27]
Mn	39.42	23.65	6.1	0.9	[26]	[27]
Ni	26.15	15.69	4.8	0.3	[26]	[27]
Ti	2.82	1.69	4.3	0.3	[31]	[27]
Other	*	*	*	*	-	-

* assumed zero

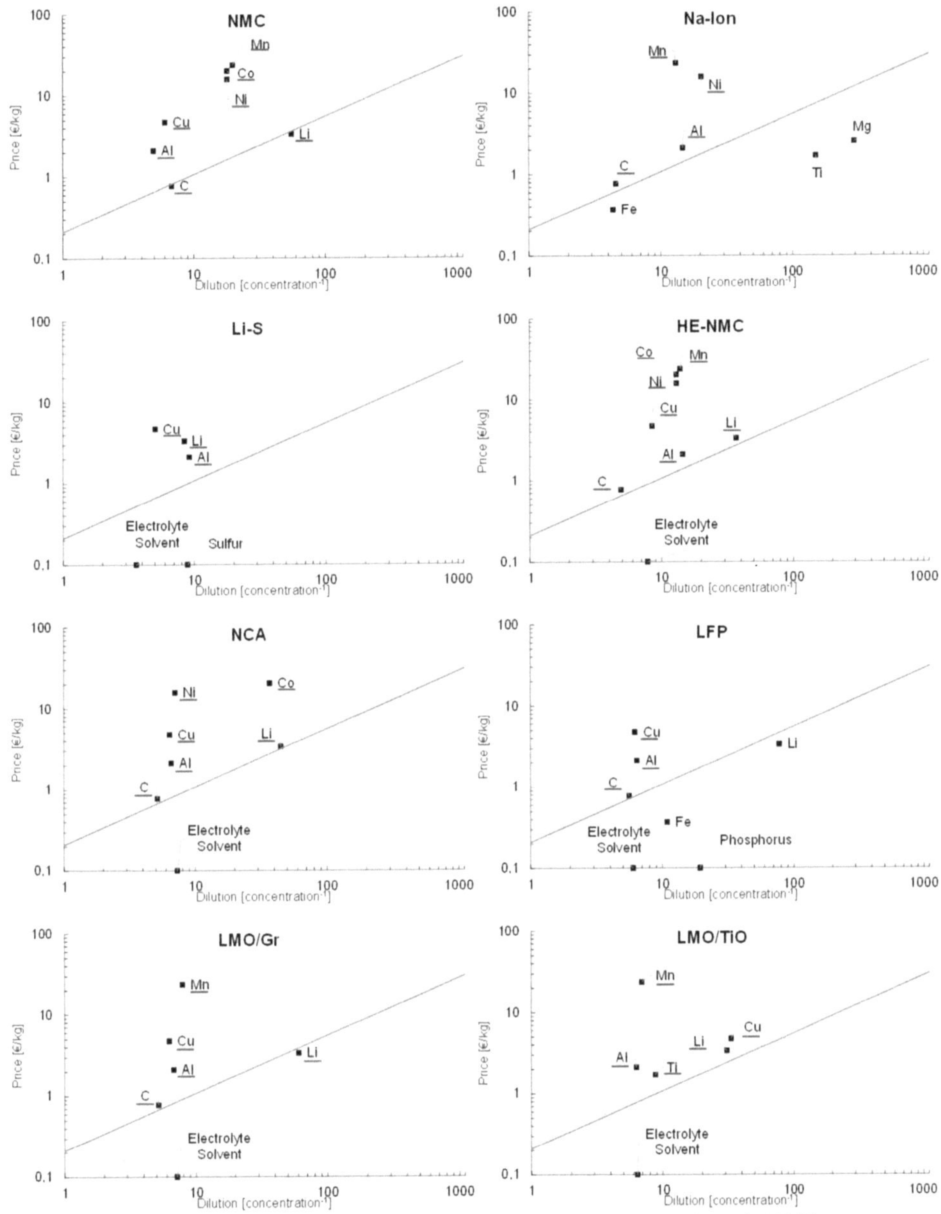

Figure 6. Sherwood plots for battery cells for NMC, Na-Ion, Li S, HE-NMC, NCA, LFP, LMO/Gr and LMO/Ti

4.4 Step 4: Calculation of Information Theory Entropy

The results from the Sherwood plots are used to determine the materials included within the Shannon entropy calculation. Not all materials are considered for the Shannon entropy. A reason is the lack of information about the materials or material compositions of some components that are missing in the inventory data provided in the studies. As shown in Table 3, the range of considered mass fraction lies between 70 % for NMC and 43 % for Li-S.

The results of the Shannon entropy method are displayed in Figure 7. In order to establish comparability between the battery technologies, the results for the material value are normalized to the battery capacity (per kWh). As the present paper does not compare different products, but instead different technologies for the same product, this approach is applied in accordance to the normalization approach by Anctil and Fthenakis [10].

The Shannon entropy ranges from the relatively simple Li-S battery technology (H = 1.18) to the complex technology of NMC batteries (H = 2.09). Both considered NMC battery packs have a relatively high Shannon entropy that relates to a high product complexity. The Shannon entropy for most traction batteries lies within a relatively narrow range of 1.5 to 2.0.

The results for the sum of single material values indicate a significant difference between the battery technologies. The material value per kWh considered for recycling ranges between 6.78 € per kWh battery capacity for Li-S to 31.37 € per kWh battery capacity for the NCA/Gr technology.

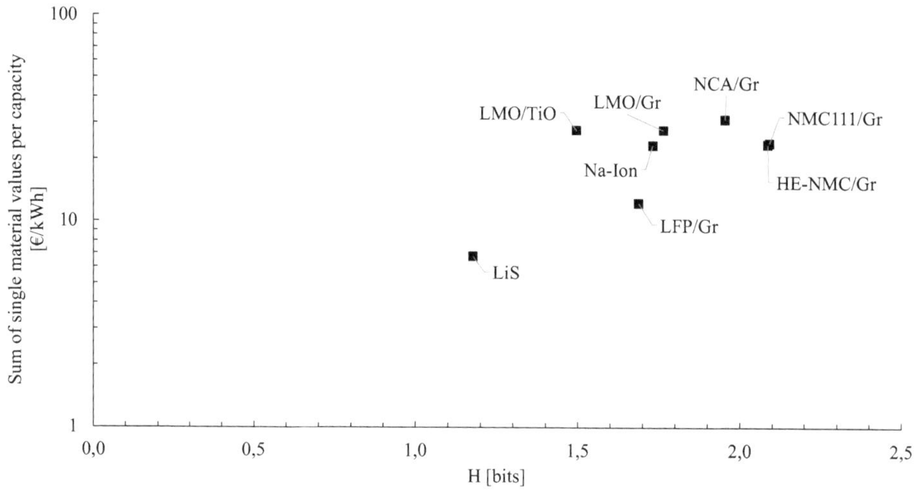

Figure 7. Shannon entropy for battery chemistries; sum of single material values are normalized to 1 kWh.

Table 3. Selected results of case study

Battery Type	**Mass fraction considered in Shannon entropy**	**Shannon entropy (H)** [bits]	**Sum of material value** [€/kWh]	**Economic Importance (EI)** [EI/kWh]	**Supply Risk (SR)** [SR/kWh]
NMC	70 %	2.09	23.96	18.14	3.83
HE-NMC	65 %	2.09	23.56	12.38	3.87
LMO/Gr	64 %	1.76	27.69	20.02	5.17
LMO/Ti	48 %	1.50	27.69	17.35	1.86
NCA/Gr	69 %	1.95	31.37	24.57	6.01
LFP/Gr	60 %	1.69	12.28	27.15	6.64
Na-Ion	64 %	1.73	23.40	25.48	7.20
Li-S	43 %	1.18	6.78	8.36	0.92

4.5 Step 5: Integration of Economic Metrics

Economic metrics considered in this study include the sum of material values embedded in the cell, the Economic Importance score (EI) and the Supply Risk (SR) of the EU. The results for EI and SR for each battery technology are displayed in Table 3. They have been normalized to the battery capacity in order to enable comparability between the battery technologies, since technologies with a lower energy density require more material or cells to reach the energy requirements for an EV battery pack. From a recycling perspective, high values for the embedded material values, EI and SR as well as low values for the Shannon entropy positively influence the recyclability.

The main contributor on the SR results is the Graphite of the anode. It contributes in-between 60% to 70% to the SR for all battery cells using graphite. Li-S and LMO/TiO achieve significantly lower SR scores for their material mixes, since they use different anode materials.

Approximately 80% of the sum of material values for recycled material for both NMC cells is contributed by Nickel, Manganese and Cobalt. Considering the drastically rising prices for these materials in recent years, the efficient recycling of these cathode materials will be crucial for the recycling industry. The high specific energy of the NMC technology leads to relatively low EI and SR scores per kWh.

Other current battery technologies, such as LFP and NCA also have relatively high EI and SR of the embedded materials. LFP cells consist of relatively cheap materials that reduce the embedded material value, whereas the high Nickel content in NCA is a main contributor to the high material value.
The material mix of Li-S cells has a relatively low value, with low values for the EI and SR. Therefore, it can be stated, that it is unlikely that they will achieve high recycling rates from a purely economic perspective. Na-Ion cells on the other hand achieve high values for EI and SR, which are mainly caused by the high Manganese content within the cells, which makes up to 50% of the value of recycled materials within the cell.
Finally, it is important to notice that the materials used for the module and cell housing, cables and power electronics also have high material values and EI respectively SR scores that makes them reasonable target materials for recycling. Therefore, an efficient pack and module disassembly is a critical step in battery recycling.

4.6 Step 6: Analysis and Discussion

The results for the selected battery technologies for this case study show a high dilution of embedded materials in different current Li-Ion batteries. Two trends will have a major impact on the dilution. First, current technology development aims at lowering the expensive and critical metals within the battery, such as Co, Mn and Li. This leads to higher dilutions and eventually worse recyclability. Second, the general aim of current development is to lower the inactive materials in order to achieve higher energy densities. One the one hand, this trend results in a positive influence on the recyclability due to the better-balanced material concentration of the battery system. Inactive materials from the housing will have a higher dilution and active materials from the cells a lower dilution. On the other hand, on a cell perspective the material dilution decreases, which makes it more difficult to separate the materials efficiently and has a negative impact on possible recycling rates.
The Sherwood plots in Figure 6 show that most materials used within the cells have a relatively high material value. In recent years, the price for materials for EV batteries has increased dramatically [6]. Higher material values have a positive influence on the recyclability and may have a critical influence on industry and governments to implement efficient recycling structures and to ensure a high return rate of spent batteries. The NMC technologies are the most complex battery technology identified in this study. NMC is expected to increase its market share significantly to up to 68% in 2025 [6]. Together with the presented technology trends, this poses a major challenge for the development of efficient recycling processes and the recyclability of battery cells.
In order to increase the recyclability of Li-Ion battery cells, two strategies can be identified through the application of the methodology. The economic value (sum of single material values) per kWh can be increased and the material mixing complexity within the cells can be reduced. Both strategies imply incorporating more expensive and scarce materials like Li, Co, Ni and Mn. This stands in contrast with the explained current trends in battery development and industry interests. However, as the analysis of the Economic Importance and the Supply Risk indicates, there is a strategic interest in securing the embedded materials for future production purposes. Hence, governmental regulations that aim to increase the recycling quotes and to strengthen a circular economy are likely to be implemented in the coming years. Furthermore, new business models for EV batteries, e.g. product service systems where the battery manufacturer remains the owner of the battery, can provide incentives to design better recyclable batteries as the companies will be able to recover more materials in a better quality.

5 Conclusion

This paper presents a six-step methodology to determine the recyclability of products by contrasting material mixing complexity and economic incentives for recycling. The methodology is applied in a case study on current and future traction battery cell technologies and establishes comparability between the technologies by normalizing the results to the battery capacity (per kWh). Generally, battery cells are relatively complex products with a high material mixing complexity. Predictions about a technology's recycling rate were not made, because most batteries are situated on the 'apparent recycling boundary' defined by Dahmus and Gutowski [5]. Nevertheless, there are significant differences regarding the determining factors for the recyclability of different battery technologies. Li-S was identified as the technology with the lowest material mixing complexity, but also with the lowest sum of recycled material value as well as the lowest scores for EU Economic Importance and Supply Risk. LMO-based technologies have moderate material mixing complexity and high scores in the economic metrics, indicating that LMO-based technologies possess a comparatively good recyclability. The same applies to Na-Ion battery cells. Whereas LMO is perceived as a declining technology, the latter is a technology with potential application in the future. NMC is the currently preferred technology with a high research intensity. The results from this study indicate its recyclability is worse than other technologies. The trend towards higher material dilution of expensive materials will further decrease the recyclability. It requires further research to fully analyse this effect.
The proposed methodology is suitable for comparing different product technologies as done in the case study. Due to the electrochemical processes throughout the battery life, the embedded materials are hard to separate at the end of life. Hence, cell disassembly is not viable in most of the cases. The presented method provides broader information into recyclability, which can be used in DfD and DfR methods in order to improve future recycling efforts. Further extension

potential can be identified to increase the informative value and decrease the uncertainty, such as incorporating environmental metrics or the integration of a scenario based analysis with different material prices as underlying variables.

6 Zusammenfassung

Während die Bedeutung von Antriebsbatterien und Elektromobilität insgesamt an Bedeutung gewinnen, hat die Frage nach der Recyclingfähigkeit der Batterien das Interesse von Politik, Industrie und Wissenschaft erregt. Ziel dieses Beitrags ist es, das Verständnis des Recyclings von Antriebsbatterien durch Anwendung des informationstheoretischen Konzepts der Entropie zu erweitern. Zu diesem Ziel werden informationstheoretische Entropieindikatoren verwendet, um die Komplexität des Materialmixes heutige und zukünftig in Elektrofahrzeugen verwendeter Batteriechemie zu bestimmen. Durch die Integration verschiedener ökonomischer Werte sowie industrieller, politischer und sozialer Einflussfaktoren wird die Recyclingfähigkeit der Antriebsbatterien evaluiert und die Enwticklung zukünftiger Recyclingsysteme für Batterien diskutiert. Die Ergebnisse zeigen, dass die verwendete Methode geeignet ist um verschiedene Produkttechnologien zu vergleichen und dass signifikante Unterschiede in Bezug auf die die Recyclingfähigkeit bestimmenden Faktoren verschiedener Batteriesysteme bestehen.

7 References

[1] Dunn JB, Gaines L, Kelly JC, James C, Gallagher KG (2015) The significance of Li-ion batteries in electric vehicle life-cycle energy and emissions and recycling's role in its reduction. Energy Environ Sci 8(3):158–168

[2] Nelson PA, Gallagher KG, Bloom ID, Dees DW (2012) Modeling the Performance and Cost of Lithium-Ion Batteries for Electric-Drive Vehicles. Technical Report. Argonne, IL (United States)

[3] Diekmann J, Hanisch C, Froböse L, Schälike G, Loellhoeffel T, Fölster AS, Kwade A (2017) Ecological Recycling of Lithium-Ion Batteries from Electric Vehicles with Focus on Mechanical Processes. J Electrochem Soc 164(1):A6184–A6191

[4] Geyer R, Kuczenski B, Zink T, Henderson A (2015) Common Misconceptions about Recycling. J Ind Ecol 20(5):1010–1017

[5] Dahmus JB, Gutowski TG (2007) What Gets Recycled: An Information Theory Based Model for Product Recycling. Environ Sci Technol 41(21):7543–7550

[6] Pillot C (2017) Worldwide Rechargeable Battery Market 2016-2025 - 2017 edition. In: Advanced Battery Power 2017. Münster, 2018

[7] Kwade A, Diekmann J (2018) Recycling of Lithium-Ion Batteries. Springer International Publishing

[8] van Schaik A (2014) Material-Centric (Aluminum and Copper) and Product-Centric (Cars, WEEE, TV, Lamps, Batteries, Catalysts) Recycling and DfR Rules. In: Handbook of Recycling, pp. 307–378

[9] Johnson J, Harper EM, Lifset R, Graedel TE (2007) Dining at the periodic table: Metals concentrations as they relate to recycling. Environ Sci Technol 41(5):1759–1765

[10] Anctil A, Fthenakis V (2013) Critical metals in strategic photovoltaic technologies: abundance versus recyclability. Prog Photovoltaics Res Appl 21:1253–1259

[11] Rechberger H, Brunner PH (2002) A New, Entropy Based Method To Support Waste and Resource Management Decisions. Environ Sci Technol 36(4):809–816

[12] Gutowski TG, Dahmus JB (2005) Mixing entropy and product recycling. Proc 2005 IEEE Int Symp Electron Environ, pp. 72–76

[13] Sherwood TK (1959) Mass transfer between phases. Priest Lect, vol. 33

[14] Allen DT, Behmanesh N (1992) Waste As Raw Materials. Ind Ecol

[15] Grübler A (1998) Technology and Global Change. Technical Report. Cambridge University Press

[16] Mohamed Sultan AA, Lou E, Mativenga PT (2017) What should be recycled: An integrated model for product recycling desirability. J Clean Prod 154:51–60

[17] European Commission (2010) Critical raw materials for the EU. Report of the Ad hoc Working Group on defining critical raw materials

[18] Schmidt T, Buchert M, Schebek L (2016) Investigation of the primary production routes of nickel and cobalt products used for Li-ion batteries. Resour Conserv Recycl 112:107–122

[19] Gaines L, Sullivan J, Burnham AJ, Belharouak I (2011) Life-Cycle Analysis for Lithium-Ion Battery Production and Recycling. In: Transportation Research Board 90th Annual Meeting. Washington DC, pp. 23–27

[20] Romare M, Dahllöf L (2017) The Life Cycle Energy Consumption and Greenhouse Gas Emissions from Lithium-Ion Batteries. Technical Report. IVL Swedish Environmental Research Institute Ltd

[21] Cerdas F, Titscher P, Bognar N, Schmuch R, Winter M, Kwade A, Herrmann C (2018) Exploring the effect of increased energy density on the environmental impacts of traction batteries: A comparison of energy optimized lithium-ion and lithium-sulfur batteries for mobility applications. Energies 11(1):150

[22] Placke T, Kloepsch R, Dühnen S, Winter M (2017) Lithium ion, lithium metal, and alternative rechargeable battery technologies: the odyssey for high energy density. J Solid State Electrochem 21(7):1939-1964

[23] Fraunhofer-Institut für System- und Innovationsforschung ISI (2015) Gesamt-Roadmap Lithium-Ionen-Batterien 2030

[24] Peters J, Buchholz D, Passerini S, Weil M (2016) Life cycle assessment of sodium-ion batteries. Energy Environ Sci 9(5):1744–1751

[25] Schönemann M (2016) Multiscale simulation approach for battery production systems. Springer International Publishing

[26] EuroStat (2017) Statistics on the production of manufactured goods Value - ANNUAL 2016. http://ec.europa.eu/eurostat/web/prodcom/data/database/

[27] European Comission (2017) Study on the review of the list of critical raw materials - Non-critical raw materials factsheets

[28] European Comission (2017) Study on the review of the list of critical raw materials - Critical Raw Materials Factsheets
[29] Statista (2018) Graphite prices worldwide from 2011 to 2020. https://www.statista.com/statistics/452304/graphite-prices-worldwide-prediction-by-flake-grade. (Accessed: 23-Apr-2018)
[30] Marscheider-Weidemann F, Langkau S, Hummen T, Erdmann L, Tercero Espinoza L (2016) Rohstoffe für Zukunftstechnologien 2016. Technical Report. DERA Rohstoffagentur, no. 28
[31] Statista (2018) Durchschnittspreise ausgewählter mineralischer Rohstoffe in den Jahren 2010 bis 2016. https://de.statista.com/statistik/daten/studie/260427/umfrage/durchschnittspreise-ausgewaehlter-mineralischer-rohstoffe. (Accessed: 23-Apr-2018)

The Importance of Recyclability for the Environmental Performance of Battery Systems

Jens F. Peters[1*], Manuel Baumann[2] and Marcel Weil[1,2]
[1] Helmholtz Institute Ulm (HIU), Karlsruhe Institute of Technology (KIT) - 76021 Karlsruhe, Germany
[*] j.peters@kit.edu
[2] Institute for Technology Assessment and Systems Analysis (ITAS), Karlsruhe Institute of Technology (KIT) - 76021 Karlsruhe, Germany

Abstract

While the global demand for batteries is increasing at a fast pace, also concerns about the impacts on resources and environment associated with their production and disposal are growing. While several studies about the environmental impacts of batteries exist, the end-of-life stage is often disregarded and the relevance of battery reuse or recycling not quantified. However, the end-of-life phase of battery storage systems is highly relevant for their overall environmental performance. In order to quantify this relevance, we extend existing LCA studies by an end-of life model and assess the influence of battery recycling for the life cycle impact of three different battery types. These include a lithium-ion battery (LIB), a vanadium-redox-flow battery (VRFB) and an aqueous hybrid ion battery (AHIB), all for stationary energy storage services (renewable support). The results show that a high recyclability can improve the environmental performance of the batteries over their life cycle significantly. The good recyclability of the VRFB can overcome the disadvantage of its lower efficiency and lower energy density, making it the best performing battery system among the assessed technologies. Similarly, the AHIB, in spite of high impacts from battery production and replacement, can become competitive if fully recycled, but only if electrode components can be reused directly without the need for breaking them down and processing them into their precursors. This underlines the need for a design for recyclability of batteries for minimising environmental impacts of battery systems and the corresponding loss of valuable resources.

1 Introduction

The transition from fossil fuels towards a renewable energy based society requires a re-structuring of the energy system. Within this, the storage of electricity from fluctuating renewable energy sources for its use on demand is one of the major challenges [1]. While numerous technologies exist for energy storage, the choice of the most appropriate one is often difficult and requires a thorough assessment of different technology options specifically for the envisaged application. Numerous studies about potential environmental impacts of energy storage systems are available, especially batteries, which are among the most promising options for short-and medium term storage of electricity [2]. However, the majority of these studies focus on the production and use phase, especially that of lithium-ion batteries (LIB), while their disposal or recycling is majorly disregarded [3]. Other studies evaluate the end-of-life (EoL) stage of LIB separately, quantifying the environmental impacts and benefits associated with different recycling processes or comparing recycling with other options for disposal of used batteries (incineration or landfilling) [4–6], while only a few include the EoL stage in their assessments [7–9]. Apart from a general uncertainty due to the lack of detailed inventory (LCI) data on recycling processes, this hinders comparing different battery technologies under a full life cycle perspective. While of lower importance when comparing very similar batteries (e.g., different LIB) with similar end-of-life processes, this becomes more severe when comparing very different approaches for electrochemical energy storage. In a first attempt towards an all-embracing assessment, we evaluate the importance of considering recycling processes for the comparison of conceptually different battery technologies for stationary energy storage.

2 Methodology

2.1 Assessment Framework

For quantifying the relevance of battery recycling for the environmental impact of battery systems, a life cycle perspective is required. Life Cycle Assessment (LCA) is a standardized method for this purpose, accounting for the compilation and evaluation of all the resources and potential environmental impacts associated with a product life cycle, from extraction through the use phase to the end-of-life handling, including recycling processes [10–12]. Consequently, the present study considers all stages of the lifecycle of the stationary battery systems within their application. The functional unit (FU) is defined as 1 MWh of electricity provided (discharged) by the battery over the 20 year lifetime of the application. This includes the production of the batteries (with all upstream processes until the extraction of raw materials), possible battery replacements due to limited lifetime of the batteries, the electricity 'lost' during charge/ discharge due to internal inefficiencies, and the end-of-life (EoL) handling of the batteries i.e., their recycling. Onshore wind and open field photovoltaic (PV) installations in average German locations are considered as electricity sources.

A. Pehlken et al. (Eds.), *Cascade Use in Technologies 2018*,
https://doi.org/10.1007/978-3-662-57886-5_13

Impact assessment is done according to the CML method, considering four impact categories: global warming potential (GWP), human toxicity potential (HTP), acidification potential (AP) and depletion of abiotic resources, minerals and metals, reserve base (ADP) [13]. The life cycle modelling and assessment is carried out using openLCA 1.6[14] and ecoinvent 3.3 (cut-off model) as background LCI database.

2.2 Battery Manufacturing

The battery production stage is modelled based on existing LCA studies [15–18]. The inventories provided by these previous works are adapted to meet the specific requirements of the assumed applications (as described in more detail in Section 2.3). Three conceptually very different battery technologies are considered, all with properties especially suitable for stationary applications and dimensioned for delivering with the same effective net storage capacity:

- An LIB (Lithium-Ion battery) : A rack-mounted stationary battery system with several air-cooled battery trays that carry 18650-type round cells with an LFP-LTO type cell chemistry optimized for longevity [15, 18]
- An AHIB (aqueous hybrid ion battery): A hybrid lithium/sodium based battery with aqueous electrolyte and a layout similar to that of conventional lead-acid batteries with comparably low energy density) [16]
- a VRFB (vanadium redox flow battery): special type of battery where the liquid electrolyte is stored in external tanks and is pumped through a stack where the electrochemical reactions take place [17]

The LIB is modelled based on a recent publication on stationary batteries [19] that assessed a 26kWh capacity battery as a rack mounted stationary system. However, the battery configuration is adapted to match the requirements of the assumed application in terms of storage capacity, 6 MWh effective net capacity (see next chapter). A battery installation comprising 271 racks is required for this purpose, with a design capacity of 7 MWh (Table 1). No further infrastructure (buildings, foundations, etc.) is accounted for.
For the AHIB model, a recent publication on this battery type is used. The underlying publication assesses a 26kWh battery module for home storage purpose, why the inventory data is scaled linearly accordingly to fit the capacity requirements. Due to lower depth of discharge and efficiency (Table 1), a design capacity of 8.43 MWh is required for providing an effective 6 MWh of capacity.
The considered VRFB system shows a rated power of 1 MW and an energy capacity of 8.3 MWh. The principal battery layout is derived from published literature, majorly a recent publication on stationary VRFB [17, 20]. The battery composition on a mass basis is estimated based on the dimensions of the different parts provided there. An 8.3 MWh system requires an electrolyte volume of 194 m^3 and, assuming a void volume of 50% and a tank capacity of 300 m^3. For a nominal power of 1 MW, two stacks with a total of 155 cells are required. Each stack is composed of a membrane, a bipolar plate, two carbon felt electrodes, plus current collectors and gaskets. Peripheral components comprise centrifugal pumps, pipes, cables and electronic components. As for the other battery types, other infrastructure components like buildings, foundations, etc. are disregarded.

Table 1. Design parameters of the stationary battery systems. Calendric lifetime refers to the lifetime of the stack (VRFB) or the battery cells (LIB, AHIB), while the electrolyte and periphery components of the VRFB are assumed not to be replaced over the lifetime of the application. SoC = State of Charge

	VRFB	**LIB**	**AHIB**	Unit
Efficiency	0.75	0.90	0.83	--
Min SoC	0.05	0.05	0.20	--
Design capacity	8.30	6.97	8.43	MWh
Effective capacity	6.00	6.00	6.00	MWh
Cycle life	10000	8000	4000	cycles
Calendric life	10	17.5	15	years
Effective energy density	19.4	32.6	12.5	Wh/kg

2.3 Battery Operation

The stationary batteries are dimensioned for providing renewables support (RS) for wind or PV installations. The assumed power capacity for this application is 1MW and the net storage capacity 6 MWh [21]. The net capacity considers the internal inefficiencies of the batteries and the minimum state of charge (SoC), requiring a certain oversizing of the batteries. For providing net 6 MWh, a nominal capacity of 8.3 MWh is required for the VRFB with the assumed operation parameters, 8.4 MWh for the AHIB and 7.0 MWh for the LIB. The application (RS) requires an average 1.12 cycles per day over 20 years, i.e. a total of 8176 charge-discharge cycles over the lifetime [2]. The two different electricity sources, wind or PV, are assumed to show identical load profiles. This is a major simplification, but allows evaluating the influence of different internal efficiencies of the batteries for electricity associated with different environmental impact under similar framework conditions. As typical for RS, the power requirements are low (1 MW power for a 7 / 8.3 MWh battery), i.e. comparably slow charge and discharge times. This favours the AHIB, but also the VRFB, which are suited especially for low power applications. The LIB could provide much higher power rates in the same battery configurations, which is why for power intensive applications like peak shaving or primary regulation, the results might differ significantly. Since the batteries are charged and discharged roughly once a day, possible self-discharge effects are neglected. For battery operation, only the electricity lost due to internal inefficiencies is considered, but not the impacts of the discharged electricity. Due to the selection of the functional unit (1MWh of electricity provided over lifetime), the amount of electricity discharged (provided to the end-user) and thus the corresponding impacts would be identical for the different batteries, which is why this can be excluded without altering the final results.

2.4 Recycling

Recycling of the assessed batteries is still in an early phase, and only very limited data can be found. Several processes exist for recycling of LIB, which recover copper and steel from the cell housing, cobalt and nickel, and eventually manganese and $LiCO_3$ (depending on the process technology) [6]. A recovery of phosphate or titanium might be technically feasible [22], but current industrial recycling processes do not aim at recovering these metals due to their low economic value. Due to the very limited data availability, a simplified approach is used for the end-of life model of the LIB, using major standard ecoinvent datasets. The battery installation is dismantled mechanically and the racks and tray housings separated from the battery cells. The energy inputs required for dismantling and compaction are approximated based on the ecoinvent process 'scrap sorting and pressing'. Steel parts, cables, waste electronic components and waste plastic are send to separate specific recyclers, using ecoinvent data for recycling of these components. The isolated battery cells are crushed in a shredder prior to recycling and then fed entirely into the recycling process. The obtained shredding product is a mixture of all cell materials and needs to be processed by hydrometallurgical or pyrometallurgical processes for separating and recovering part of them: the hydrometallurgical process recovers $LiCO_3$, copper and iron, while the pyrometallurgical process recovers only copper and iron [4, 5, 23]. Both processes are designed for recycling of mixed streams of waste LIB, thus they are not optimised towards recovering the materials contained in LFP-LTO cells (phosphorous and titanium). This represents the current state of the art; however, future recycling processes optimised for this specific battery chemistry might achieve higher recovery rates.
A general recycling efficiency of 90% is assumed for the basic metals steel and copper and for lithium in the hydrometallurgical route, while the remaining fraction is sent to waste treatment by incineration. In contrast to the LIB, the AHIB and VRFB show a low level of integration and the major part of the components can be dismantled manually and all parts may be recycled separately and on a macro-scale, eventually even recovering the individual cell components or the active materials in pure form. The energy input required for disassembly is comparably low and approximated with the ecoinvent process 'iron scrap sorting and pressing'. For the AHIB, the active material pellets are recovered directly and can then be refurbished or recycled. No information exists about possible recycling options for the AHIB electrodes, which is why two options are considered: direct re-use for new batteries without further processing (optimistic) or disposal as waste (pessimistic). Some re-processing will certainly be necessary due to degradation effects in the active materials, which is why the truth is probably in between these two extremes. For the VRFB, the electrolyte does not degrade significantly, but requires some re-processing and purification. For this purpose, the electricity input required for re-balancing the electrolyte is accounted for, estimated based on electrochemical calculations (amount of electricity required for bringing the electrolyte fractions from a homogeneous mixture to the required oxidation states) [24]. All remaining components of the dismantled batteries are sent to specific treatment, modelled with the corresponding ecoinvent processes for recycling of metals, plastics and electronic components.
According to the ecoinvent cut-off model, the environmental impacts associated with the recycling process are attributed totally to the batteries, while the recovered materials are considered free of burden. For quantifying the potential benefit of recycling, two options need to be compared therefore: (i) using the average market mix for each of the required materials (majorly virgin materials) and (ii) internal recycling using the materials recovered from battery recycling (burden free according to cut-off system model) in the share determined by the assumed recycling efficiency. Comparing these two allows quantifying the benefit that can be obtained from battery recycling.

3 Results and Discussion

Figure 1 provides the characterisation results for the assessed battery systems with and without recycling. Without recycling (left four panels of Figure 1), the AHIB obtains the worst results in three of the four assessed categories. This is a result of the low energy density and thus the high mass of battery required for providing a given amount of storage capacity. Additionally, the AHIB has a limited lifetime and requires complete replacement during the assumed 20 year lifetime of the application, which roughly doubles the impacts from battery manufacturing. Within the battery manufacturing stage, one of the key contributors is the tetrafluoroethylene-based binder, whose production is associated with high environmental impacts. Substituting this by an alternative fluorine-free binder would certainly be a significant improvement. For AP, the AHIB and the VRFB perform similar. The comparably high impact of the VRFB is due to the vanadium mining which is associated with significant SO_2 emissions from mining and beneficiation. The LIB battery in general shows good results, also due to its long lifetime and high internal efficiency, which minimises the impacts arising from the use phase. This is especially relevant for electricity with higher associated impacts, as can be observed when comparing the results for wind and PV electricity. For more carbon intensive electricity like e.g., grid electricity with a certain share based on fossil energy, the relevance of the use phase would increase even further.

When including the potential benefit due to the use of recycled material in the manufacturing of the batteries (four panels on the right side of Figure 1), the picture changes. Impacts from battery manufacturing and -replacement are reduced, while the use phase now plays a much more prominent role. This can be observed especially for the VRFB with its relatively low internal efficiency and when charging electricity from PV installations, where the use-phase is by far the most dominating contributor. The impacts from the end-of-life processing are comparably small for all battery types. However, it has to be considered that the modelling of the recycling processes is a very simple approximation with high uncertainty and might therefore underestimate the associated impacts.

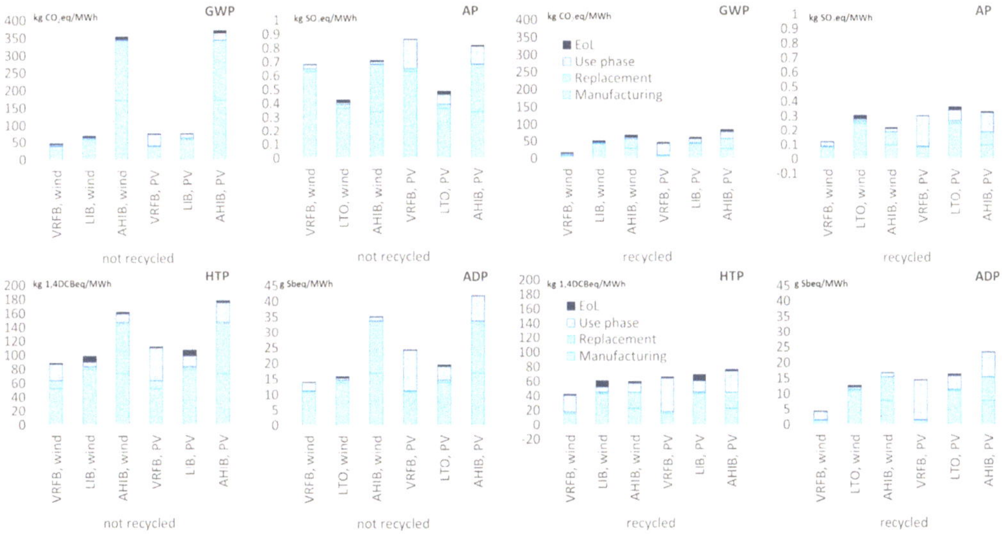

Figure 1. Environmental impacts per MWh of electricity delivered by the assessed battery systems over lifetime, charging electricity from wind or PV installations. Left four figures without recycling (using virgin raw materials for battery production), right four figures with recycling (recycled material substitutes the corresponding amount of virgin raw material). GWP = Global warming potential; ADP = abiotic depletion potential, metals and minerals; AP = acidification potential; EP = eutrophication potential; HTP = human toxicity potential; ODP = ozone depletion potential; POF = photogenic ozone formation potential

Recycling reduces the environmental impacts associated with the battery systems significantly. As shown in Figure 2, producing batteries with material from recycled batteries (circular material flow in the production process) reduces the impacts from the production stage between 15 and 90%. The assessed batteries show different recyclability, with the highest reduction (up to 90 %) obtained by the AHIB, while recycling of the LIB battery shows lower environmental benefits. As previously mentioned, the AHIB can easily be dismantled by mechanical processes on a macro-scale, recovering the individual cell components as discrete parts, while the highly integrated LIB require higher inputs for processing and less material can be recovered, reducing the potential benefits of recycling.

However, it has to be taken into account that very favourable assumptions are made for the AHIB, considering a direct re-use of the recovered electrode pellets without the need for separating binder, aggregates and active material and the re-processing of the active material. In reality, the active materials might be degraded more severely, requiring the processing into pre-cursor materials and the synthesis of new electrode pellets from these. While this recycling process reduces the amount of raw materials needed, it still requires new binder materials, new additives and a new sintering of the electrode pellets. Since the binder is one of the main contributors to GWP, but also to other categories, this would affect the outcome strongly. No information is available about what an AHIB recycling would look like, thus a more precise modelling of the corresponding processes and efficiencies is not possible. In order to provide an idea of the possible range of impact reduction that AHIB recycling might achieve, Figure 3 shows the environmental impact associated with the provision of 1MWh of electricity from the AHIB system for the two extreme assumptions:

(i) direct re-use of the recovered electrode components ('electr. re-use'; optimistic assumption), and
(ii) disposal of the electrodes and use of virgin materials for the manufacture of new electrode pellets, while all remaining battery components are recycled as assumed previously ('electr. disposal'; pessimistic assumption).

While the electrodes would certainly not be disposed of completely in the recycling process, these two extremes (direct re-use of the electrodes without further need for refurbishment and disposal and replacement by new ones) indicate a range, with the actual recycling benefit somewhere in between. The differences are significant, indicating that the recycling of the electrode pellets is key to reducing the environmental impacts of this battery type. Taking into account that the binder is one of the key contributors to the total impact of the AHIB, a recycling process recovering also the binder, or even a process for re-furbishing the electrode pellets while maintaining their integrity should therefore be aimed at.

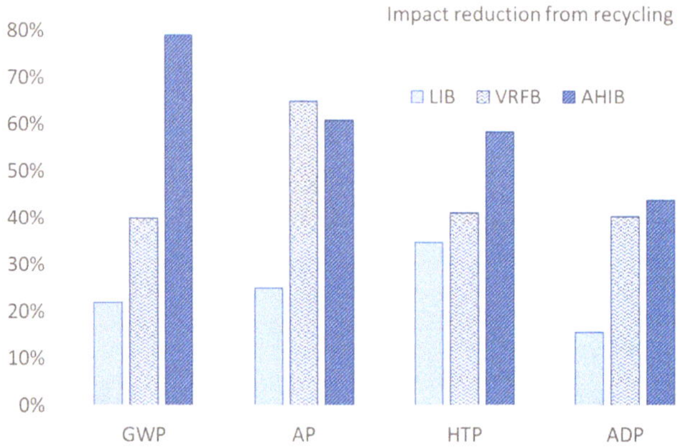

Figure 2. Reduction of impacts per MWh of electricity provided by the storage system due to the use of recycled materials

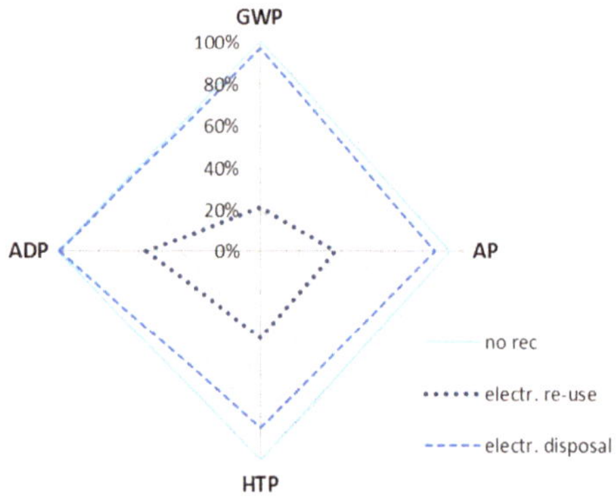

Figure 3. Impacts of the AHIB relative to the worst scoring battery in each category, without recycling, with direct re-use of the active material pellets / electrodes ('component rec') and with re-processing of the active materials for recovering raw materials ('material rec').

While highly integrated batteries like the LIB have advantages under a cradle-to-gate perspective (higher energy density and thus lower material demand per provided capacity, but also higher charge-discharge efficiency), less integrated systems can have significant advantages when it comes to recyclability. The AHIB and VRFB are easy to dismantle and all major components can be recovered by mechanical dismantling on a macro-scale. The highly integrated LIB require complex processes and obtain a commingled fraction of micro-size particles that are difficult to separate and require significant process inputs while only recovering a fraction of the materials originally contained in the batteries. This can change the picture towards an advantage of technologies easy to dismantle on a macro-scale (AHIB and VRFB), especially for electricity sourced majorly from renewable sources. For electricity with a higher share of fossil energy sources, a high internal efficiency can be more important for the overall environmental performance. Thus, different battery technologies have to be assessed case-specifically, taking into account the individual requirements of a given storage application. However, design for recyclability can be considered highly important in terms of a future circular economy and might even outweigh the possibly reduced energy density or lower performance.

4 Conclusions

Considering that the end-of-life stage does affect the outcomes of the assessment significantly, cradle-to-gate approaches might often be inappropriate for comparing different battery technologies, even though the use-phase is included. Thorough modelling of the end-of-life phase is essential for obtaining meaningful results, especially when comparing very different storage technologies. In this sense, an urgent need for more detailed and robust inventory data for battery end-of-life processes can be pointed out, especially for recycling processes. The results from the preliminary assessments show that design for recyclability is very important in the sense of a circular economy and the benefits from good recyclability might easily outweigh the possibly reduced energy density or lower performance of a battery designed for easy recycling. However, not all recycling is alike, and the recovery of entire components for direct re-use can obtain much higher benefits than the recovery of individual elements / pre-cursor materials from a commingled product from e.g., a pyrometallurgical process. In this sense, considering aspects of recyclability already in early development stages can help to maximise recycling efficiency and to reduce material demand and environmental impacts of battery systems significantly. Especially for comparing very different battery technologies as in the present study, a good recyclability can be the decisive aspect for turning a certain technology into the environmentally most favourable one.

5 Zusammenfassung

Während die globale Nachfrage nach Batterien stetig ansteigt, nehmen ebenso die Bedenken bezüglich des durch Produktion und Entsorgung verursachten Einflusses auf Ressourcen und Umwelt zu. Zwar behandeln einige Studien die Umweltauswirkungen von Batterien, jedoch wird häufig die End-of-Life Phase des Lebenszyklusses vernachlässigt und die Bedeutung von Wiederverwendung bzw. Recycling der Batterien nicht quantifiziert. Die End-of-Life Phase der Batteriespeichersysteme ist jedoch hochrelevant für ihre übergreifende Umweltleistung. Um diese Bedeutung zu quantifizieren, werden in diesem Beitrag bestehende LCA-Studien durch ein End-of-Life Modell erweitert und der Einfluss des Batterierecyling auf den Life Cycle Impact dreier Batterietypen bestimmt. Diese beinhalten eine Lithium-Ionen-Batterie (LIB), Vanadium-Redox-Durchflussbatterie (VRFB) und eine Aqueous-Hybrid-Ionenbatterie (AHIB), die alle jeweils zur stationären Energespeicherung verwebdet werden. Die Ergebnisse zeigen, dass eine hohe Recyclingfähigkeit die Umweltleistung der Batterien über den gesamten Lebenszyklus signifikant verbessern kann. Die hohe Recyclingfähigkeit der VRFB kann deren Nachteile einer geringeren Effizienz und geringerer Energiedichte ausgleichen, so dass dieses Batteriesystem unter den hier verglichenen am besten abschneidet. Zudem kann auch die AHIB bei vollständigem Recycling trotz hoher Umweltbelastungen in Produktion und Austausch wettbewerbsfähig werden, jedoch nur wenn die Elektrodenkomponenten direkt wiederverwendet werden und nicht in ihre Vorstufen zerlegt werden müssen. Dies unterstreicht die Notwendigkeit eines „Design for Recyclability" der Batterien, um die Umweltauswirkungen sowie den Verlust wertvoller Ressourcen zu minimieren.

Acknowledgments. The authors thank the Helmholtz Excellence-Network „post-Lithium Batteries" for funding.

6 References

[1] Agora Energiewende (2014) Stromspeicher in der Energiewende. Agora Energiewende, Berlin, Germany

[2] Baumann MJ, Peters JF, Weil M, Grunwald A (2017) CO2 footprint and life cycle costs of electrochemical energy storage for stationary grid applications. Energy Technol 5:1071–1083 . doi: 10.1002/ente.201600622

[3] Peters JF, Baumann MJ, Zimmermann B, Braun J, Weil M (2017) The environmental impact of Li-Ion batteries and the role of key parameters – A review. Renew Sustain Energy Rev 67:491–506

[4] Buchert M, Jenseit W, Merz C, Schüler D (2011) Ökobilanz zum „Recycling von Lithium-Ionen-Batterien" (LithoRec). Öko-Institut, Darmstadt, Germany

[5] Buchert M, Jenseit W, Merz C, Schüler D (2011) Entwicklung eines realisierbaren Recyclingkonzepts für die Hochleistungsbatterien zukünftiger Elektrofahrzeuge – LiBRi. Teilprojekt: LCA der Recyclingverfahren. Öko-Institut

[6] Fisher K, Wallén E, Laenen PP, Collins M (2006) Battery Waste Management Life Cycle Assessment

[7] Dunn JB, Gaines L, Sullivan J, Wang MQ (2012) Impact of Recycling on Cradle-to-Gate Energy Consumption and Greenhouse Gas Emissions of Automotive Lithium-Ion Batteries. Environ Sci Technol 46:12704–12710 . doi: 10.1021/es302420z

[8] Dewulf J, Van der Vorst G, Denturck K, Van Langenhove H, Ghyoot W, Tytgat J, Vandeputte K (2010) Recycling rechargeable lithium ion batteries: Critical analysis of natural resource savings. Resour Conserv Recycl 54:229–234 . doi: 10.1016/j.resconrec.2009.08.004

[9] Unterreiner L, Jülch V, Reith S (2016) Recycling of Battery Technologies – Ecological Impact Analysis Using Life Cycle Assessment (LCA). Energy Procedia 99:229–234 . doi: 10.1016/j.egypro.2016.10.113

[10] EC-JRC (2010) ILCD Handbook: General Guide for Life Cycle Assessment - Detailed guidance. European Commission - Joint Research Centre. Institute for Environment and Sustainability, Ispra, Italy: EC-JRC - Institute for Environment and Sustainability

[11] ISO (2006) ISO 14040 – Environmental management – Life Cycle Assessment – Principles and framework. International Organization for Standardization, Geneva, Switzerland

[12] ISO (2006) ISO 14044 – Environmental management – Life Cycle Assessment – Requirements and guidelines. International Organization for Standardization, Geneva, Switzerland

[13] Guinée JB, Gorrée M, Heijungs R, Huppes G, Kleijn R, Koning A de, Oers L van, Wegener Sleeswijk A, Suh S, Udo de Haes HA, Bruijn H de, Duin R van, Huijbregts MAJ (2002) Handbook on life cycle assessment. Operational guide to the ISO standards. I: LCA in perspective. IIa: Guide. IIb: Operational annex. III: Scientific background. Kluwer Academic Publishers, Dordrecht

[14] GreenDelta GmbH (2017) OpenLCA 1.6. http://www.openlca.org/. Accessed 5 Sep 2017

[15] Peters JF, Weil M (2018) Providing a common base for life cycle assessments of Li-Ion batteries. J Clean Prod 704–713

[16] Peters JF, Weil M (2017) Aqueous hybrid ion batteries- an environmentally friendly alternative for stationary energy storage? J Power Sources 364:258–265

[17] Weber S, Peters J, Baumann MJ, Weil M (2018) Life Cycle Assessment of a Vanadium-Redox-Flow Battery. Environ Sci Technol submitted:

[18] Bauer C (2010) Ökobilanz von Lithium-Ionen Batterien. Paul Scherrer Institut, Labor für Energiesystem-Analysen (LEA), Villigen, Switzerland

[19] Peters JF, Weil M (2017) Aqueous hybrid ion batteries – An environmentally friendly alternative for stationary energy storage? J Power Sources 364:258–265 . doi: 10.1016/j.jpowsour.2017.08.041

[20] Minke C (2016) Techno-ökonomische Modellierung und Bewertung von stationären Vanadium-Redox-Flow-Batterien im industriellen Maßstab, 40th ed. Cuvillier Verlag, Göttingen

[21] Baumann MJ, Peters JF, Weil M, Grunwald A (2017) CO2 footprint and life cycle costs of electrochemical energy storage for stationary grid applications. Energy Technol submitted:

[22] Li H, Xing S, Liu Y, Li F, Guo H, Kuang G (2017) Recovery of Lithium, Iron, and Phosphorus from Spent LiFePO4 Batteries Using Stoichiometric Sulfuric Acid Leaching System. ACS Sustain Chem Eng 5:8017–8024 . doi: 10.1021/acssuschemeng.7b01594

[23] Fisher K, Wallén E, Laenen PP, Collins M (2006) Battery Waste Management Life Cycle Assessment. Environmental Resources Management (ERM)

[24] Roznyatovskaya N, Herr T, Küttinger M, Fühl M, Noack J, Pinkwart K, Tübke J (2016) Detection of capacity imbalance in vanadium electrolyte and its electrochemical regeneration for all-vanadium redox-flow batteries. J Power Sources 302:79–83 . doi: 10.1016/j.jpowsour.2015.10.021

Assessment of the Demand for Critical Raw Materials for the Implementation of Fuel Cells for Stationary and Mobile Applications

Rikka Wittstock[1], Alexandra Pehlken[2], Fernando Peñaherrera[2], Michael Wark[3]
[1]Department of Accounting and Information Systems, University of Osnabrueck, Osnabrueck, 49074, Germany
[2]Cascade Use, Carl von Ossietzky University of Oldenburg, Oldenburg, 26111, Germany
[3]Department of Chemistry, Carl von Ossietzky University of Oldenburg, Oldenburg, 26129, Germany
alexandra.pehlken@uol.de

Abstract

Because of their low emissions and possible contribution to sustainable development, both mobile and stationary fuel cells show promising tendencies to play an important role in the future. However, the polymer exchange membrane fuel cell (PEMFC) used in two major applications (i.e. fuel cell vehicles (mobile application) and household Combined Heat and Power systems (stationary application)) contains significant amounts of platinum, a material considered critical within the European Union due to its geological scarcity and highly concentrated supply base. Using material flow analysis, this paper seeks to examine how the implementation of mobile and stationary fuel cells will affect demand for critical raw materials and to what degree recycling presents a viable option for reducing the pressure on primary production. Based on a number of developed scenarios, it is demonstrated that neither the platinum requirements arising from a more widespread adoption of fuel cell vehicles nor the platinum demand from household heating systems is likely to cause a depletion of platinum deposits in the near future. However, both technologies may increase the pressure on the already constricted platinum market, thus rendering this resource even more critical to European industries.

1 Introduction

Both mobile fuel cells in automotive applications and stationary fuel cells in Combined Heat and Power (CHP) household heating systems have not reached market maturity yet, but show promising tendencies to play an important role in the future if these technologies are supported by adequate political decisions. The main driver for both technologies is clearly their low production of emissions during the use phase.

While fuel cell vehicles represent the mobile application of fuel cells (FC), the stationary fuel cells addressed in this paper represent household heating systems only.

In general, the distributed energy generation from stationary fuel cells (SFC) offers significant benefits. With electrical efficiency values of 33% up to 60% under experimental conditions for power-only systems, as well as a combined efficiency in cogeneration higher than 90% [1], SFC exhibit high energy conversion efficiencies, which represents primary energy savings and a reduction of transmission losses. Using natural gas as fuel, SFC can reduce CO_2 emissions due to their efficient conversion of natural gas. Even combustion of natural gas consisting to > 95 % of methane emits per calorific value (in J) about 30% less CO_2 than gasoline, about 60 % less than bituminous coal and even more than 90% less than lignite. In CHP units, fuel cells present a solution to cut building energy use and emissions in the near term, as the technology can make use of existing fuel distribution infrastructure [2].

Unlike battery electric vehicles, fuel cell vehicles are comparable to internal combustion engine vehicles with regards to driving range and performance and can be considered the lowest-carbon option for medium to long-distance trips. Since this segment represents 75% of today's passenger vehicle CO_2 emissions, substitution of one combustion engine vehicle with a fuel cell vehicle (FCV) achieves a comparably higher CO_2 reduction [3].

In developed regions, particularly in Germany, Japan, South Korea and the United States, the fuel cell industry has gained traction in recent years [4].

However, the catalyst material platinum, which is generally applied in the polymer exchange membrane fuel cell (PEMFC) used for both mobile and stationary fuel cells, has attracted the world market's attention since it was classified as highly critical as part of the European Commission's assessment of critical raw materials [5]. In addition to the significant environmental impacts caused by the extraction of platinum group metals (PGM), including emissions of sulphur dioxide, of CO_2 equivalents in the range of 13,000 tons per ton of PGM [6], excessive water and energy consumption [7], habitat destruction, air and water pollution and generation of dust, particulate matter and solid waste [8], platinum is considered a critical metal in the EU due to its geological scarcity, its use in a variety of technologies and its highly concentrated supply base [5].

Metal recycling is considered an important strategy for lowering the pressure of primary deposits and the environmental impacts of extraction, as well as increasing the economic benefit of the metals in use [9]. The recycling of platinum is already well-established and can therefore act to fill gaps within the primary supply chain of platinum. This internal ability for supplying secondary raw materials from recycling can serve to make the European Union (EU) more independent of other continents' suppliers. However, recycling rates vary significantly between different applications. While valuable precious metals, including platinum, typically have high recovery rates, the recycling rates for certain

A. Pehlken et al. (Eds.), *Cascade Use in Technologies 2018*,
https://doi.org/10.1007/978-3-662-57886-5_14

consumer applications are disturbingly low. Passenger cars represent one such application with relatively low platinum recycling rates, as the recycling quotas for exhaust gas catalysts reach a mere 50% to 60% [10]. This makes the automotive industry the largest net consumer of platinum today, even when growth of vehicle sales is ignored [11].
It is therefore in the interests of both fuel cell and car manufacturers to close these loops and focus on the contribution of recycling for meeting their future platinum demand. The same holds true for the newly established technology of fuel cells in household CHP units, for which no collection and recycling is taking place so far since very few fuel cell CHP units have reached the end of their useful lifetime as of 2018. In addition to PEMFC in stationary application also Solid Oxide fuel cells have entered the market in household application, where platinum does not play any role. However, it is not known which technology is preferred by which company and the data is very poor. In order to assess supply issues of the main catalyst material used in both technologies and thus explore the sustainability of these alternative sources of energy, this paper seeks to examine how the implementation of mobile and stationary fuel cells will affect demand for critical raw materials and to what degree recycling presents a viable option for reducing the pressure on primary production.
The paper is structured as follows: Section 2 provides an overview of the state of the art for fuel cells in both stationary and mobile applications. A material flow assessment of the relevant fuel cell materials, with a strong focus on platinum is applied in Section 3, while Section 4 concludes with a discussion of the results.

2 Theoretical Background

2.1 Characteristics of Fuel Cells

From a structural perspective, fuel cell vehicles can be considered a type of hybrid vehicle, in which the fuel cell replaces the internal combustion engine [12]. Using atmospheric oxygen and compressed gaseous hydrogen supplied from the onboard tank, the fuel cell generates electricity, which powers the vehicle's electric motor. Due to their favorable attributes, such as low operating temperature, fast start-up and fast response to varying loads, FC installed in automotive applications are currently only of the PEMFC type (status 2017).
A more detailed description of the structure and functioning is provided by, e.g. [13]. In summary, in PEMFC a platinum-based catalyst is generally used for both the oxidation reaction taking place at the anode and the reduction reaction occurring at the cathode. Particles of the catalyst (10 to 100 nm in size) are finely dispersed on a porous substrate, which usually consists of high surface area carbon powders, e.g., carbon black. Commercial production today typically employs procedures like those of the printing industry, in which the supported or unsupported catalyst material is mixed with solvents, binder (perfluorosulfonic acid or other ionomers in protonated form) and other additives to form a "catalyst ink" and then applied in wet form [13-16].
Assuming a fuel cell stack of 80 kWe, the platinum loading reported for 2013 still translates into a total platinum requirement of around 20 g per fuel cell vehicle. Moreover, as these are fuel cell stacks produced and tested under laboratory conditions, the platinum load of fuel cell stacks installed in FCV technology currently 'on the road' is likely still higher. Due to the proprietary nature of such data, exact and reliable numbers for the platinum content of FCV models manufactured today cannot be obtained; instead, assessment of the current platinum load relies on estimates and proposals published by automobile manufacturers and researchers [13].
For fuel cell-generated energy supply, two fuel cell types are mostly used. The most developed technologies for stationary applications up to 10 kWe are the PEMFC and the Solid Oxide Fuel Cell (SOFC); for fuel cells in mobile applications, mainly automotive, the PEMFCs dominate.
Based on thermodynamics, PEMFCs can reach efficiencies in the conversion of chemical energy (fuel) to mechanical energies at the wheels up to 95%, whereas classical combustion energies are limited to about 40%, due to the unavoidable Carnot process conditions. They have great technological potential as their development is already advanced due to their similarity to the ones used in automotive applications. The PEMFC for domestic energy supply is a low-temperature version with an operating temperature in the range of 80 °C, while the high-temperature mobile version can oscillate between 160 to 180 °C. PEMFC are also attractive to the residential CHP market because of their relatively compact size and the fact that they do not require insulation. PEMFC systems are to run with pure hydrogen. Since hydrogen is currently mostly obtained from natural gas by steam reforming or other fossil fuels by partial oxidation, carbon monoxide (CO) has to be removed first, since CO acts as an effective poison for the platinum catalyst due to strong adsorption. This adsorption is especially problematic at low operation temperatures below about 140°C. Thus, the hydrogen fuel has to be carefully reformed before entering the system [17].
As with PEMFC in automotive applications, a platinum catalyst is generally used for both the oxidation reaction taking place at the anode and the reduction reaction occurring at the cathode. Although the platinum content is considered one of the main cost drivers of PEMFC [13], the amount of platinum required in SFC is even higher than in mobile FC. This is due to the continuous requirement on operations, which necessitates loadings in the range of 0,75 g platinum/kWe for stationary applications in comparison to 0,2 g platinum/kWe for mobile ones [18]. The catalyst metal influences not only the specific power, but also the fuel cell's lifetime.
In contrast to PEMFCs, SOFCs operate at a temperature level of 650 to 850 ° C. In this temperature range, according to thermodynamics, fuel cells do not work more efficiently than gas-fired combustion engines, however, they can be very

easily build as CHP units. Compared to PEMFC, SOFC have the advantage of an easier gas treatment since CO adsorption is no longer possible at this high temperature [19]. SOFC are expected to be more economical in sizes above 1 MW (Brown et al. 2007, p. 2176). While the SOFC does not contain platinum, other critical materials provoke equal concerns regarding security of supply, price stability and sustainability.

Yttria (yttrium oxide) is a critical material commonly used in SOFC membranes. The yttrium oxide is used as a dopant (about 8 wt.-%) in the zirconia (ZrO_2) in order to ensure the necessary conductivity of oxide (O^{2-}) anions. While zirconium can be considered as an element under observation with respect to the availability, yttrium is a rare-earth element (REE), and is categorized as a critical metal due to its low availability (around 9 000 t/a of production), and the concentration of its reserves. The problem of recycling of REE including yttrium is that in applications they are mostly used in low concentrations and in strong interaction with other elements. Recycling of these mixed compounds by the common route of employing redox-active ligands is often difficult since the chemical coordination properties of the different elements might be similar [20]. For recycling via electrochemical separation typically the thermodynamic rules the process. Since this is not useful for REEs, here the kinetics must be used to control the separation [21]. Thus, e.g. low temperatures might be favourable. Other potentially problematic materials include scandium, lanthanum and cobalt.

Although the SOFC technology is not addressed further in this paper, the issues concerning critical materials such as Yttrium should be examined once the market for SOFC demonstrates a higher maturity.

2.2 Market Application of Fuel Cells

Stationary Fuel Cells (SFC) refer to units designed to provide power at a fixed location. They include small and large stationary power supply, backup or uninterruptible power supply, combined heat and power, and combined cooling and power. SFC can be applied in various stationary applications, ranging from systems under 1 kWe for CHP, larger units (several kWe) for district heating or large buildings, up to MWe applications for industrial cogeneration and electricity production without cogeneration. Different conventional systems are already well established for each of these applications, such as gas engine CHP, gas turbines, or combined cycle power plants [22].

They can be broadly divided into two types: power-only and cogeneration systems. The power-only type includes back-up or uninterruptible power systems fuelled by hydrogen [23]. Cogeneration systems use the electric output and the waste heat for heating applications. With around 665 TWh [24], the energy used by the 18.9 million German households [25] accounts for around 26.2% of the total final energy use. The majority of CHP systems for domestic energy supply (micro-CHP) in the market correspond to the following technologies: Otto Motor, Stirling Motor, Steam Expansion, Organic Rankine Cycle (ORC), Micro-Gas turbine, and SFC [26].

Replacing conventional gas boiler systems, natural gas fuelled micro-CHP systems are expected to become one of the first major mass markets for FC. This product can be described as a mass-produced appliance targeted at residences (single and multi-family), and small commercial users [27]. The power produced is preferentially used within the building, with top-up and backup power supplied from the grid. In addition, any excess power produced is also exported to the grid. The size of domestic application systems typically ranges between 500 We and 5 kWe, with small commercial units reaching up to 50 kWe [27, 28]. For industrial-sized applications, SFC systems can reach ranges of 200 to 500 kWe [28].

Deployment of the system is oriented towards maximization of the efficiency [18], i.e. the energy output. Compared with electricity generation in central power plants and heat generation, SFC demonstrate a 30% better utilization of the fuel [27]. Further advantages include short startup times, a high cycle stability and good partial load capacity, making them suitable for intermittent and load-variable operation. Disadvantages, on the other hand, include high purity of the required fuel, for which a technically complex gas purification process is necessary [29].

In contrast to SFC [30], the mobile FCs have already entered the market and are beginning to compete with conventional combustion engines vehicles and other alternative powertrain technologies. The limited availability of hydrogen fuelling stations continues to be one of the greatest obstacles for market penetration of FCV. The expansion of the hydrogen station network is largely taking place in certain lead markets, including California (which has 33 hydrogen fueling stations serving approximately 4,200 FCV), Germany (45 hydrogen stations) and Japan (91 hydrogen stations) [31]. In Europe, two different FCV models are currently available for purchase, although this number is likely to expand as other manufacturers enter the market [32]. Future development of the FCV market is likely to depend both on private investment decisions and on the political endorsement of FCV technology.

3 Methodology of Material Flow Analysis

Recovering metals from complex and low-grade materials can be more energy intensive than supplying them from raw ores. In many cases, however, the recovery of secondary metals is less energy intensive than mining, as the metal concentration in many products is higher than in ores. This is the case, for example, for precious and rare metals in electronic products [33]. In order to examine not only the demand for critical metals arising from the implementation of stationary and mobile fuel cells, but also the viability of meeting a portion of this demand through recycling, it is therefore vital to assess the amount of recoverable metals accumulated in end-of-life products. As method for assessing the amount of materials we chose Material Flow Assessment (MFA), since it is based on the principle of conservation of matter or mass balancing (i.e. input equals output). It is closely linked to the concept 'metabolism of the anthroposphere', which can be understood as an analogy to the human or natural metabolism. While a more detailed review of this idea is beyond the scope of this paper, suffice it to note that the concept imitates the continuous cycling of material and energy present in biological processes and has been used to analyse and describe urban and regional material balances, especially for the purpose of environmental protection and waste management. The dynamic MFA for this paper is performed using the software tool STAN (short for SubSTance flow ANalysis), which is a freeware developed by the Institute for Water Quality, Resources and Waste Management at the Vienna University of Technology in cooperation with INKA software.

3.1 Material Flow Analysis for Automotive Fuel Cells

The model of platinum flows (Figure 1) arising from a further market penetration of fuel cell vehicles is developed applying a top-down approach: Figures for the future vehicle stock are derived from the two conservative scenarios for market penetration developed by the European Commission [34]. This report was selected from a number of studies because it covers the most far-reaching time horizon (2010 to 2050) and for the equally important reason that the contained data is easily accessible. Based on sociological, technological and economic analyses as well as extensive interaction between science and industry experts from a range of member states, the study produced market penetration scenarios for various hydrogen-based technologies, including four on FCV. In doing so, the study focuses on the following ten European countries, which hence also form the geographical focus of the MFA presented in this paper: Finland, France, Germany, Greece, Italy, the Netherland, Norway, Poland, Spain and the United Kingdom. Of these, Greece was later excluded in this study due to a lack of data on vehicle ownership. Given the current state of the FCV, the assumptions underlying the two more conservative scenarios appeared closer to reality.

In addition to the market penetration scenarios described above, certain assumptions have been made concerning technical aspects. These include the current and future platinum load per kWe installed power, material leakage during the FCV use phase, as well as production and assembly of FC and FCV and the respective efficiencies of these processes. The calculation originates from the FCV market penetration ratios given by the two scenarios in [34] which are applied to the analysed EU member states' extrapolated vehicle stock to derive the absolute number of FCV in use per annum over the considered time scale of 35 years. In the next step, the annual energy requirements (kW/a) are calculated from the number of newly registered FCV per annum. Applying to this the platinum load per kW of the respective year as well as any losses occurring in the production stage, one can then determine the gross platinum requirements per annum.

In order to determine the net platinum requirements (Figure 2), the potential supply from recycling is then calculated. Based on two recycling scenarios (i.e. "Baseline", applying efficiencies currently reported in literature, and "Pro-Recycling", assuming a highly efficient recycling chain), any platinum losses throughout the use phase and within the recycling chain are subtracted from the potentially available platinum content of end-of-life FCV assuming a vehicle lifetime of 10 years.

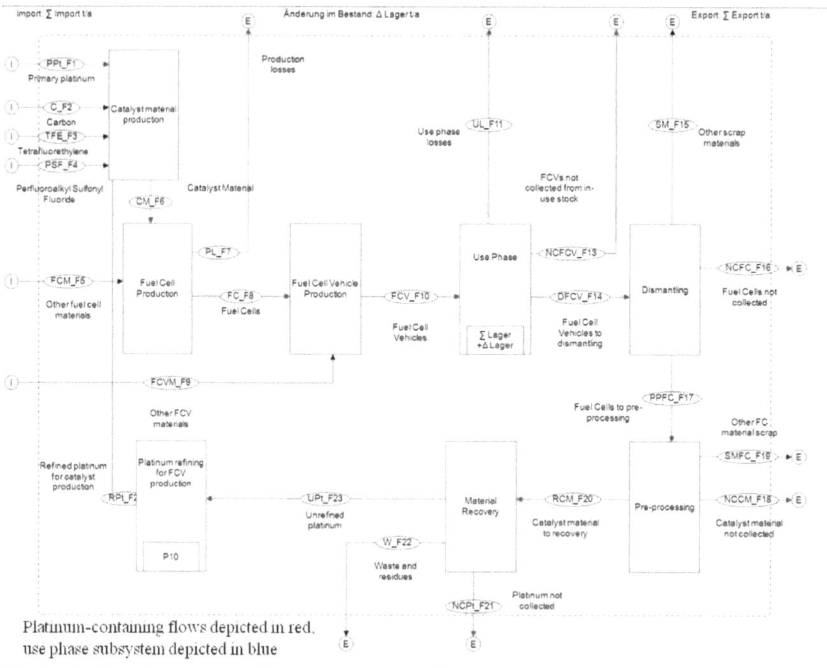

Figure 1. STAN Material Flow Analysis model for automotive fuel cells

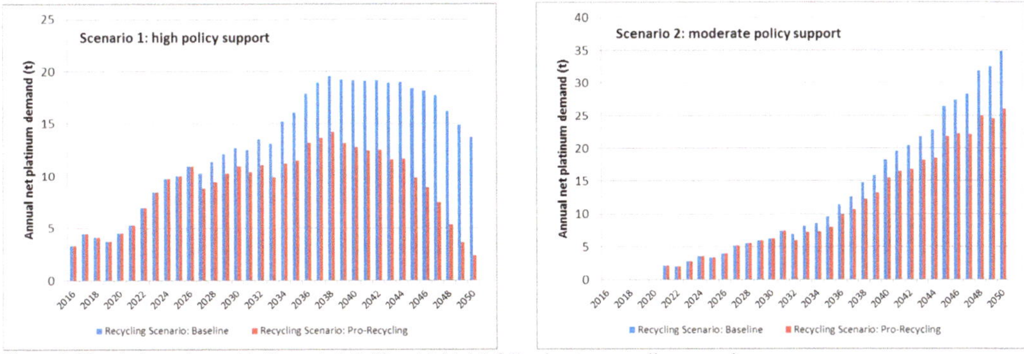

Figure 2. Net platinum demand in Scenario 1 (left) and 2 (right) following two recycling scenarios

From these calculations, the cumulative gross demand for platinum, the cumulative amount of platinum recovered through recycling and the amount of platinum lost through recycling inefficiencies can be determined. These results are summarized in Table 1. The various material efficiency ratios of production use phase and recycling are used as transfer coefficients and transferred into the MFA software, together with the values obtained from the aforementioned calculations. A detailed description of the market penetration scenarios, the two recycling scenarios and the specific calculations can be reproduced by conferring to [13].

Table 1. Cumulative demand and amount of platinum recovered per scenario (adapted from [13])

	Scenario 1 (high policy support)	Scenario 2 (moderate policy support)
Cumulative gross platinum demand	537.06 t	459.24 t
Cumulative amount of platinum recovered (Baseline scenario)	84.8 t (208.71 t lost)	44.85 t (110.38 t lost)
Cumulative amount of platinum recovered (Pro-Recycling scenario)	230.02 t (63.49 t lost)	119.17 t (36.06 t lost)

Both the annual platinum requirements and the cumulative platinum requirements shown above indicate that the diffusion of FCV according to either scenario will have a significant impact on global platinum demand. At this point in time, it is impossible to establish in how far any platinum demand from the fuel cell industry will be additional to or replace the demand from other areas of application, such as that of automotive exhaust gas catalysts. The results of Table 1 also demonstrate that the diffusion of FCV will not cause the depletion of platinum resources, as the calculated 537.06 t and 459.24 t cumulative platinum requirements are far below the currently estimated reserves of 66,000 t of platinum group metal reserves, even though any contribution from the recycling of FCV is not yet included in this extrapolation. At the same time, it must be emphasised that these figures consider only nine European lead markets. A similar level of market penetration in other parts of the world could therefore increase the required quantities of platinum dramatically.

In contrast, a comparison of the maximum annual net demand established by the two scenarios with the current global supply shows that the future platinum requirements for European FCV production only could place a significant strain on the global platinum market. With peaks of 19.5 t and 26.06 t, respectively, the European FCV industry alone would require a maximum of 12% and 16% of the 161.74 t produced in 2014. This could not only lead to significant price increases, but also raises the possibility of supply shortages as well as the dependency of a number of industries on a critical resource and emphasises the importance of recycling.

With regards to the role of recycling in meeting the platinum requirements of a growing FCV fleet, Figure 2 and Table 1 show a significant recycling potential, the exploitation of which could greatly reduce this industry's dependence on the volatile platinum market. Nonetheless, in both Scenario (1) and (2) the discrepancy between the two recycling scenarios becomes apparent. Considering the entire time span from 2016 to 2050, Table 1 lists the cumulative platinum demand for FCV as well as the shares recovered under the four scenario combinations. While the platinum losses are substantial even in the Pro-Recycling scenario, the losses documented as part of the Baseline scenario are excessively high.

3.2 Material Flow Analysis for Stationary Fuel Cells

In addition to automotive fuel cells, a model for Material Flow Analysis on PEMFC stationary fuel cells is constructed to analyse the demand for platinum, considering the different recycling scenarios for end-of-life (EoL) products (Figure 4). The model is divided into different stages: Raw Material Acquisition, Production, Use, and EoL. For each process, transfer coefficients (TF) are defined as fractions that are either lost from the system or are reincorporated to the EoL recycling process. Values for the transfer coefficients (as depicted in Table 3) are taken from literature sources when existing, or assumed by the authors after comparing them to similar technologies, under consideration of the uncertainties that come with these values. 15% of the catalyst material is lost in the electrode coating process, with the ratio of process losses improving continuously. Process losses in the region of 5% to 20% can be expected for automated, industrial-scale coating of membrane rolls, with a catalyst ink scrap rate of 10% [13]. Fraunhofer IPA, as quoted by [35], suggests the typical overspray losses for different coating methods depends on the specific method used as well as the workpiece structure, and can be as high as 90%. Additional losses can occur during the process of fabrication of the fuel cell stack, as a fraction of the manufactured PEMFC CHP units may not comply with quality control procedures. Assuming a high value of confidence for the manufacturing process, the fraction of discarded products is usually lower than 2% [36, 37] with target values of 0,1%.

The technical life time of the stack for future technological configurations is required to be at least 6 years to be commercially applicable [38], with lifetimes assuming 5 000 h/a of operation for intermittent residential operation. Calendar lifetimes would be lower for commercial and industrial installations with longer running hours [39]. The amount of PGM lost in the exhaust of a fuel cell system over its operation is insignificant [40]. [41] indicates use phase material losses of 0,35%/a, and values of 0,68%/a are used by [13] for mobile FC. Due to the continuous operation in stationary applications, this value is also applied.

Recovery rates for PEMFC for automotive applications are estimated to be greater than 95%. The high technical recyclability of PGM means that over 95% recovery can be achieved once platinum-containing products reach a state-of-the-art refining facility [10]. An overall global recycling quota for platinum of 70% (currently about 45%) should be the minimum target for 2020 [42], and 80% for 2030 [43, 44]. For SFC no recycling has taken place so far, but a high recovery rate as in mobile application is expected. Because of the need for having a heating expert in order to change the heating system, a loss of end-of-life systems by both legal and illegal export and recycling activities is unlikely.

Since stationary fuel cells have entered the market only recently and data availability, especially with regards to the use phase, is very poor, the MFA was performed considering only household installations within Germany until 2050. Therefore, this analysis of the housing market in Germany is based on estimations for the number of residential units to be constructed, replaced or renovated, as these units are a potential market for new SFC CHP units. Replacement of conventional operating systems for SFC CHP units is not deemed as economically feasible, hence only new units are considered. The different scenarios assume that SFC run on natural gas or syngas, so existing infrastructure is used [28].

Almost a third of the energy demand in Germany corresponds to housing and buildings. A portion of this is used for warm water and space heating. More than 17 million residential heating devices in Germany produce heat by burning gas and oil [19].. More than three quarters of the residential buildings are heated by central heating. About 15.1 million of these devices are installed in One- or Two-family houses, while 2.3 million gas and oil-burning devices are installed in multi-family houses [26]. One-family houses (OFH), small multi-family houses (SMH), and large multi-family houses (LMH) with 7 to 12 units , as well as large residential buildings (RBH) with more than 13 units, are the principal potential markets for SFC [29]. The number of buildings and its annual evolution corresponds to the sales market and the potential number of CHP plants.

Natural gas is used in 47.8% of the OFH and SMH, and in 47.7% of the larger residential buildings [25]. Micro CHP units with 1 to 2 kW_e and 3 to 10 kW_{th} power based on the Otto and Stirling engine and SFC technologies provide this energy [26]. A residence with 4 people demands an average of 5 MWh_e/a. This represents 570 W of average power. Most of the household applications will be smaller than 1 kW [19].

The market for new CHP units is related to the number of new buildings, units required as replacement for demolished units, and renovated units in which the insulation and heating systems are upgraded. Yearly rates are used to estimate these units. [29] presents two different scenarios for this development. In the Reference Scenario, the expansion rates are fixed at 0.5%/a. In the Ecological Scenario, the expansion rates are continuously growing and are set so that by 2050, all buildings existing before 1998 have been either replaced or renovated. To estimate the future market potential of SFC, this technology is assumed to achieve a market share of 90% within the CHP household applications market by 2050 under all scenarios [46]. The application of these values results in the number of buildings requiring a new heating system. Considering the fraction of residential buildings that use natural gas (44.7%) the potential market for FC CHP systems is calculated using forecasted rates for entrance of FC technology. [29] proposes increasing penetration rates of the technology into the market. Two scenarios are used for modelling: Scenario A (Ecological Scenario), of fast introduction of technology, and Scenario B (Reference Scenario), of slow introduction of technology. Both scenarios still consider a technology testing period. The number of FC CHP units for each residence type is calculated based on the above given distributions.

Performing simulations, [29] already determined the optimal size of a SFC CHP system for economic feasibility. Two business models for use of the SFC CHP units are considered: Private User and Contracting model. Each model results in different optimal sizes ranging from units smaller than 1 kWe to larger than 10 kWe (Table 2). RBH are not considered due to their small number.

Table 2. Optimal installation size of FC CHP units. Data from [29]

Building Type	Private User			Contracting Model		
	Financially feasible size		Operation Hours	Financially feasible size		Operation Hours
	[kW_e]	[kW_{th}]	[h]	[kW_e]	[kW_{th}]	[h]
OFH	0.5	0.7	5 700	1.8	3.6	4 400
SFH	2.0	2.9	4 800	6.0	8.6	5 500
LFH	5.0	7.1	6 500	11.0	15.7	5 200

The values of operation hours and the models for failure rates are used to calculate the number of SFC CHP units that need to be placed into the market, as units for new systems or as replacement for existing systems. The Private User model is applied to the reference scenario, while the Contracting Model is applied to the Ecological Scenario. The results indicate a growing market for SFC, with close to 800 thousand units required by 2050 under the highest demand scenario. Although the number of houses with CHP systems varies, an extended operating lifetime for the Ecological Scenario reduces the demand. The total amount of SFC can be quantified in kWe of peak power placed in the market, which is obtained multiplying the number of units and the individual optimal power (Figure 3).

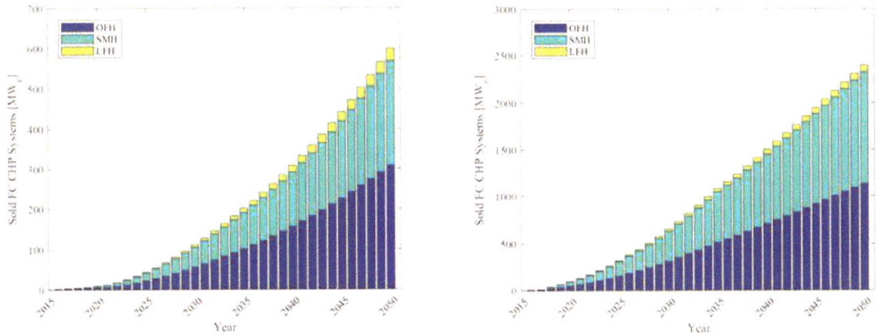

a) Reference Scenario b) Ecological Scenario

Figure 3. FC CHP units placed in the market, in MWe [47]

The value of platinum content in units coming in and out of the service phase is used in a MFA model (Figure 4) to calculate the required input of raw material. The model considers losses in different production stages, and different recovery rates for EoL stages. The values from the transfer coefficients (TC) for each step in the MFA Model are taken from literature sources when existing or assumed by comparing them to similar technologies. A model is built in Simulink 9.0, which allows further analysis of dynamic scenarios and uncertainties.

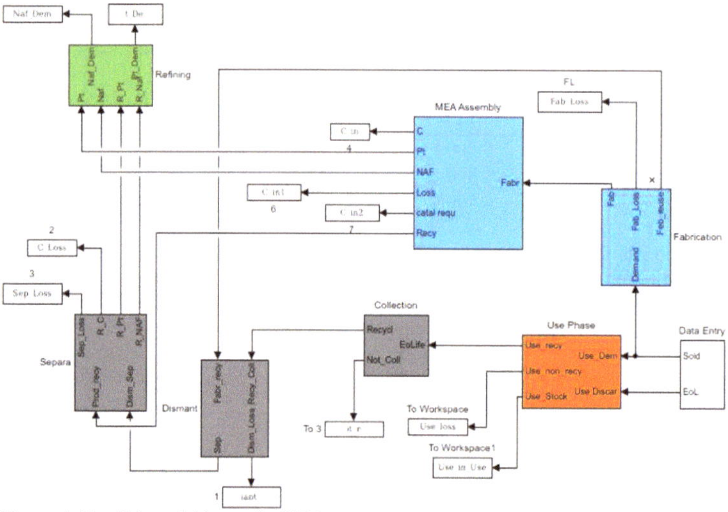

Figure 4. Simulink model layout for MFA

The Reference Scenario assumes values for the transfer coefficient on the lower spectrum of the values found in the literature review, and assumes these values remain constant, i.e. without improvement. The Ecological Scenario assumes higher values and a continuous improvement in the recycling and collection transfer coefficient, and a reduction in production and fabrication losses.

Table 3. Transfer coefficients used for MFA

Value	Reference Scenario	Ecological scenario	
		Start (2017)	Target (2050)
MEA production loss	40%	15%	5%
FC production loss	50%	2%	0.1%
Recollection rate	15%	60%	95%
Dismantling loss	3.4%	3.4%	1%
Separation loss	10%	5%	1%

A maximum of 1.3 tons of platinum for the Reference Scenario and 0.18 tons of platinum in the Ecological Scenario are estimated to be required annually as raw material input for the supply of PEMFC units for until 2050 for heating applications in German households (Figure 5). Due to improvements in recycling and stabilization of the demand, the ecological scenario presents a further reduction of annual raw platinum demand. In contrast, the input of platinum in the manufacturing stage reaches 3 tons of platinum for the Reference Scenario and 2 tons for the Ecological Scenario. The difference between the demand for raw material and material input for manufacturing are due to improvements in recycling.

a) Reference Scenario

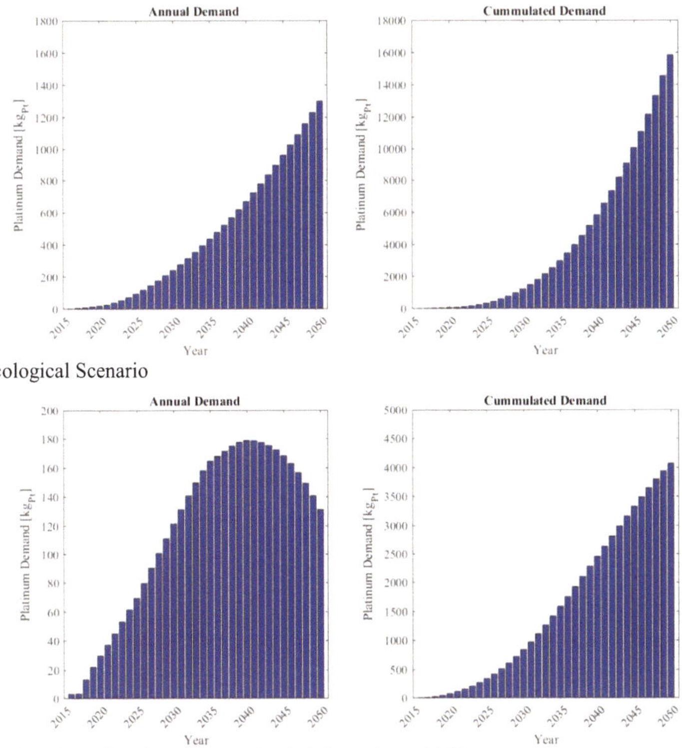

Figure 5. Values of yearly and accumulated platinum demand [47]

Due to the lack of data available for the life cycle of a SFC for heating applications in general, the values applied as transfer coefficients are subject to uncertainties that may affect the results. As of 2018, there is very little statistical data available and more field studies are needed to assess the market for FC until 2050.

4 Conclusion

Aiming to examine how the implementation of mobile and stationary fuel cells will affect demand for critical raw materials and to what degree recycling presents a viable option for reducing the pressure on primary production, material flow analyses assessing the flows of platinum throughout the life cycles of both fuel cell vehicles and household CHP systems were conducted. As demonstrated, based on the developed scenarios neither the platinum requirements arising from a more widespread adoption of fuel cell vehicles nor the platinum demand from SFC household heating systems is likely to cause a depletion of PGM deposits in the near future. However, both technologies may increase the pressure on the already constricted platinum market, thus rendering this resource even more critical to European industries. While this effect is likely not felt from a wider adoption of SFC household CHP systems before 2050, significant additional demand for platinum from the adoption of fuel cell vehicles may arise within the coming decade. With regards to recycling, both technologies profit from the high recovery rates achievable for PGM provided the end-of-life products reach a state-of-the-art refining facility. The end-of-life supply chain that ensures PGM-containing end-of-life products arrive at a recycling facility can be considered relatively strong for SFC household heating applications. This is due to the fact that expert knowledge is required to install and uninstall the technology. Mobile fuel cell applications, such as fuel cell vehicles, however, have the drawback that markets exist outside of Europe as well and the products can easily be removed from the countries in which recycling facilities exist, causing a loss of recyclable material.

While only PEMFC were considered in this study, the SOFC technology is also of interest. A recycling chain does not yet exist for this technology, although recyclability is generally assumed to be high. Since yttrium oxide is used in the membranes of SOFC, further studies are necessary assessing the possible demand for this equally critical material as well as possibilities for recovering yttrium from end-of-life products.

5 Zusammenfassung

Aufgrund der geringen Emissionen in der Nutzungsphase zeigen sowohl mobile als auch stationäre Anwendungsbereiche für Brennstoffzellen vielversprechende Tendenzen, zukünftig an Relevanz zu gewinnen. Die für zwei Hauptanwendungsbereiche (Brennstoffzellenfahrzeuge und Blockheizkraftwerke) verwendeten Protonenaustauschmembran-Brennstoffzellen enthalten jedoch signifikante Mengen an Platin, welches von der Europäischen Union als kritischer Rohstoff eingestuft wird. Anhand einer Materialflussanalyse wird in diesem Beitrag untersucht, wie sich eine Einführung mobiler und stationärer Brennstoffzellenanwendungen auf den Bedarf an kritischen Rohstoffen auswirkt und inwiefern Recycling eine geeignete Strategie darstellt, um den Druck auf Primärquellen zu mindern. Es werden verschiedene Szenarien entwickelt, durch die gezeigt werden kann, dass eine großflächigere Einführung von Brennstoffzellenfahrzeugen und mit Brennstoffzellen betriebener Haushaltsblockheizkraftwerke in näherer Zukunft nicht zu einer Erschöpfung der Platinressourcen führen wird. Beide Technologien könnten jedoch den Druck auf den bereits jetzt eingeschränkten Platinmarkt erhöhen und so die Gefahr von Versorgungsengpässen für die europäische Industrie verstärken.

Acknowledgments. This work was possible through funding the research group Cascade Use by the BMBF (FKZ 01LN1310A) and the Netonia BMBF project (FKZ 01DS16011B).

6 References

[1] Edel M (2017) Callux, Praxistest Brennstoffzelle fürs Eigenheim. https://www.now-gmbh.de/content/2-bundesfoerderung-wasserstoff-und-brennstoffzelle/5-strom-und-waerme-mit-brennstoffzellen/1-callux/callux_p8_edel_15-11-26.pdf (Accessed 27 Nov 2017).

[2] Fuel Cell Today (2017), www.technology.matthey.com/pdf/273-274-pmr-oct12.pdf (Accessed 27 Nov 2017)

[3] McKinsey & Company (2010) A portfolio of power-trains for Europe: A fact-based analysis. The role of Battery-Electric Vehicles, Plug-In Hybrids and Fuel Cell Electric Vehicles. http://www.hydrogen.energy.gov/analysis_repository/project.cfm/PID=266 (Accessed 27 Oct 2015)

[4] Ammermann H, Hoff P, Atanasiu M, Ayllor J, Kaufmmann M, Tisler O (2015) Advancing Europe's energy systems. Stationary fuel cells in distributed generation. A study for the Fuel Cells and Hydrogen Joint Undertaking. Publications Office of the European Union, Luxemburg, Luxemburg.

[5] European Commission (2017) Communication from the Commission to the European Parliament, the Council, the European Economic and Social Committee and the Committee of the Regions on the 2017 list of Critical Raw Materials for the EU (COM (2017) 490 final). European Commission, Brussels, Belgium.

[6] Saurat M, Bringezu S (2008) Platinum Group Metal Flows of Europe, Part 1. In: Journal of Industrial Ecology, vol 12, pp. 754 - 767. doi: 10.1111/j.1530-9290.2008.00087.

[7] Mudd G (2012) Sustainability Reporting and the Platinum Group Metals: A global Mining industry Leader? In: Platinum Group Metal Review, vol 56, pp. 2-19. doi:10.1595/147106711x614713.

[8] Cairncross E (2014) Health and environmental impacts of platinum mining: Report from South Africa. Presentation on behalf of PHM. http://www.thejournalist.org.za/wp-content/uploads/2014/09/Environmental-health-impacts-of-platinum-mining1.pdf (Accessed 18 Mar 2018)

[9] UNEP (2012) Responsible Resource Management for a Sustainable World: Findings from the International Resource Panel. http://www.unep.org/resourcepanel/-Default.aspx?tabid=104289 (Accessed 27 Oct 2015)

[10] Hagelüken C (2012) Recycling the Platinum Group Metals. A European Perspective. In: Platinum Metals Review 56 (1), pp. 29-35. doi: 10.1595/147106712X611733

[11] Bernhard W, Riederle S, Yoon M (2014) Fuel Cells - A realistic alternative for zero emission. https://www.rolandberger.com/publications/publication_pdf/roland_berger_fuel_cells_20140113.pdf (Accessed 27 Oct 2015)

[12] Chan CC, Bouscayro, A, Chen K (2010) Electric, Hybrid, and Fuel-Cell Vehicles. Architectures and Modeling. In: IEEE Trans. Veh. Technol. Vol 59 (2), pp. 589-598. doi: 10.1109/TVT.2009.2033605.

[13] Wittstock R, Pehlken A, Wark M (2016) Challenges in Automotive Fuel Cells Recycling. In: Recycling, vol 1, pp. 343-364. doi: 10.3390/recycling1030343.

[14] Kaz, T (2008) Herstellung und Charakterisierung von Membran-Elektroden-Einheiten für Niedertemperatur Brennstoffzellen. Ph.D. Thesis. University of Stuttgart, Stuttgart, Germany

[15] Koraishy B, Meyers JM, Wood, KL (2009) Manufacturing of Membrane Electrode Assemblies for Fuel Cells. http://www.sutd.edu.sg/cmsresource/idc/papers/2009_Manufacturing_of_membrane_electrode_assemblies_for_fuel_cells.pdf (Accessed 9 April 2015)

[16] Simons A, Bauer C (2015) A life-cycle perspective on automotive fuel cells. In: *Applied Energy, vol* 157, pp. 884-896. doi: 10.1016/j.apenergy.2015.02.049.

[17] Brown J, Hendry C, Harborne P (2007) An emerging market in fuel cells? Residential combined heat and power in four countries. In: Energy Policy, vol 35 (4), pp. 2173-2186. doi: 10.1016/j.enpol.2006.07.002.

[18] Stahl H, Bauknecht D, Hermann A, Jenseit W, Köhler A (2016): Ableitung von Recycling- und Umweltanforderungen und Strategien zur Vermeidung von Versorgungsrisiken bei innovativen Energiespeichern. Umweltbundesamt, Dessau-Roßlau, Germany. https://www.umweltbundesamt.de/sites/default/files/medien/378/publikationen/-texte_07_2016_ableitung_von_recycling-und_umweltanforderungen.pdf (Accessed 27 Nov 2017).

[19] Töpler J, Lehmann J (2014) Wasserstoff und Brennstoffzellen. Technologien und Marktperspektiven. Springer, Berlin, Germany.

[20] Bogart JA, Cole BE, Boreen MA, Lippincott CA, Manor BC, Carroll PJ, Schelter EJ (2016) Accomplishing simple, solubility-based separations of rare earth elements with complexes bearing size-sensitive molecular apertures. In: PNAS vol. 113, pp. 14887–14892.

[21] Fang H, Cole BE, Qiao Y, Bogart JA, Cheisson T, Manor BC, Carroll PJ, Schelter EJ (2016) Electro-kinetic Separation of Rare Earth Elements Using a Redox-Active Ligand. In: Angew. Chem. Int. Ed. vol 56, pp. 13450 –13454.

[22] Pehnt M (2003) Life cycle analysis of fuel cell system components. In: Vielstich W, Lamm A, Gasteiger H (2003) Handbook of fuel cells – Fundamentals, Technology and Applications, vol 3. J.Wiley, Weinheim, Germany.

[23] Larminie J, Dicks A (2003) Fuel Cell Systems Explained. Second Edition. Wiley, Hoboken, USA.

[24] UBA (2018) Energieverbrauch nach Energieträgern, Sektoren und Anwendungen. https://www.umweltbundesamt.de/daten/energie/energieverbrauch-nach-energietraegern-sektoren (Accessed 03 May 2018)

[25] BDEW (2015) Wie heizt Deutschland? BDEW-Studie zum Heizungsmarkt. BDEW Bundesverband der Energie- und Wasserwirtschaft e.V., Berlin, Germany.

[26] Erdmann V (2013) Mikro-Kraft-Wärme-Kopplungsanlagen. Status und Perspektiven. Verein Deutscher Ingenieure (VDI). Düsseldorf, Germany. https://www.vdi.de/uploads/media/Statusreport-MKWK_2013.pdf (Accessed 27 Nov 2017)

[27] ASUE (2016) Brennstoffzellen für die Hausenergieversorgung. Funktionsweise, Entwicklung und Marktübersicht. ASUE e.V., Berlin, Germany. http://www.asue.de/sites/default/files/asue/themen/brennstoffzellen/2016/broschueren/05_03_16_asue_brennstoffzellen_hausenergieversorgung.pdf (Accessed 27 Nov 2017)

[28] HFP (2005) European Hydrogen and Fuel Cell Technology Platform. Deployment Strategy. http://www.fch.europa.eu/sites/default/files/documents/hfp_ds_report_aug2005.pdf (Accessed 27 Nov 2017)

[29] Jungbluth C (2007) Kraft-Wärme-Kopplung mit Brennstoffzellen in Wohngebäuden im zukünftigen Energiesystem, https://www.deutsche-digitale-bibliothek.de/binary/6YZTN5LJN33ASZ6TEQRU72RZSAKBRCZZ/full/1.pdf (Accessed 27 Nov 2017)

[30] Marscheider-Weidemann F, Langkau S, Hummen T, Erdmann L, Tercero Espinoza L, Angerer G, Marwede M, Benecke S (2016) Rohstoffe für Zukunftstechnologien 2016 – DERA Rohstoffinformationen 28. DERA, Berlin, Germany

[31] Green Car Reports, https://www.greencarreports.com/news/1115396_germanys-hydrogen-stations-exceed-us-california-beats-japan-on-density (Accessed 07 Jun 2018)

[32] Next Green Car, http://www.nextgreencar.com/fuelcellcars/ (Accessed 07 Jun 2018)

[33] Hagelüken C, Lee-Shin J, Carpentier A, Heron C (2016) The EU Circular Economy and its Relevance to Metal Recycling. In: Recycling 1, pp. 242-253. doi: 10.3390/recycling1020242.

[34] European Commission (2008) HyWays - The European Hydrogen Roadmap. Project Report, http://www.hyways.de/docs/Brochures_and_Flyers/HyWays_Roadmap_-FINAL_22FEB2008.pdf (Accessed 10 Aug 2015)

[35] BUBW (2015) Nasslackieren. Spritzlackieren, http://www.bubw.de/?lvl=465 (Accessed 10 Sep 2015)

[36] Matathil A, Ganapathi K, Ramachandran K (2012) Reduction of Scrap in an Electronic Assembly Line Using DMAIC Approach. In: SASTECH Journal vol 11, pp. 53-59.

[37] Shokri A, Nabhani F, Bradley G (2015) Reducing the scrap rate in an electronic manufacturing SME through Lean Six Sigma methodology, https://core.ac.uk/download-/pdf/46520216.pdf (Accessed 30 Nov 2017)

[38] DLR (2004) Final report on technical data, costs and life cycle inventories of fuel cells. Deliverable no. 9.2 - RS 1a, http://www.needs-project.org/2009/Deliverables/RS1a%20D9.2%20Final%20report%20on%20fuel%20cells.pdf (Accessed 27 Nov 2017)

[39] Dodds P, Staffell I, Hawkes A, Li F, Grünewald P, McDowall W, Ekins P (2015) Hydrogen and fuel cell technologies for heating. A review. In; International Journal of Hydrogen Energy 40, pp. 2065-2083. doi: 10.1016/j.ijhydene.2014.11.059.

[40] Grot S, Grot W (2007) Platinum Recycling Technology Development. Excerpt from 2007 DOE Hydrogen Program Annual Progress Report. Ion Power, Inc. New Castle, USA, https://www.hydrogen.energy.gov/pdfs/progress05/vii_e_1_grot.pdf (Accessed 28 Nov 2017)

[41] Kromer M, Joseck F, Rhodes T, Guernsey M, Marcinkoski J (2009) Evaluation of a platinum leasing program for fuel cell vehicles. In: International Journal of Hydrogen Energy 34 (19), pp. 8276-8288. doi: 10.1016/j.ijhydene.2009.06.052.

[42] Umicore (2005) Materials flow of platinum group metals. Umicore, London, United Kingdom

[43] Buchert M, Jenseit W, Dittrich S, Hacker F (2011) Ressourceneffizienz und ressourcenpolitische Aspekte des Systems Elektromobilität. Öko-Institut e.V., Darmstadt, Germany

[44] UNEP (2009) Critical Metals for Future Sustainable Technologies and their Recycling Potentials, http://www.unep.fr/shared/publications/pdf/DTIx1202xPA-Critical%20Metals%20and%20their%20Recycling%20Potential.pdf (Accessed 28 Nov 2017)

[45] Krewitt W, Schlomann B (2006) Externe Kosten der Stromerzeugung aus erneuerbaren
Energien im Vergleich zur Stromerzeugung aus fossilen Energieträgern: Gutachten im für das Bundesministerium für Umwelt, Naturschutz und Reaktorsicherheit, https://www.pvaustria.at/wp-content/uploads/5-erneuerbaren-Energien-fossilen-Energietraegern.pdf (Accessed 30 Nov 2017)

[46] Peñaherrera F, Pehlken A (not dated) Assessment of the Demand of Critical Materials for
Fuel Cell Micro CHP for Households in Germany, submitted to Global NEST Journal, 2018

The Material Use of Perovskite Solar Cells

Juan Camillo Gomez[1], Thomas Vogt[1] and Urte Brand[1]
Institute of Networked Energy Systems, DLR, Oldenburg, 26129, Germany,
Juan.GomezTrillos@dlr.de

Abstract

This work quantifies, through material flow analysis, the demand and discard of lead and indium in a scenario of future adoption of perovskite solar cells, considering four aspects for the construction of scenarios. The first aspect was the type of perovskite solar, which was considered either as single junction or as tandem with silicon solar cell. The second aspect considered was the future market share, with assumptions of 20% and 80% in 2050. The third aspect was lifetime of the modules, considered to be either 5 years or 30 years. Finally, scenarios with and without recycling were evaluated. The results show that the demand for lead might not be significant compared to the current supply. On the other hand, the use of indium in a high market share might go beyond the current supply of this material. The use of tandem technologies might decrease the use of materials because of higher power conversion efficiency. Finally, a longer lifetime and the recycling at the end of life might decrease considerably the amount of materials that are dispersed into the environment.

1 Introduction

Zero and low-carbon energy supplies, including renewable energy, have been acknowledged as a main part of the strategy to mitigate climate change and limit the increase of temperature in the surface of the Earth because of anthropogenic greenhouse gas emissions [1]. Among the renewable energy technologies, photovoltaic devices can supply the required electricity and at the same time diminish the emissions per unit of energy produced compared to conventional conversion using fossil fuels [2]. Multiple photovoltaic technologies have been developed to increase the energy conversion and minimize the costs, making photovoltaic technologies competitive regarding conventional fossil fuels. Perovskite Solar Cells (PSC) have been acknowledged as a photovoltaic technology with a high potential, with an increase in power conversion efficiency (PCE) from 3.8% in 2009 to 22.7% in 2017 [3]. Such PCE is now comparable to the best research efficiencies for commercial technologies such as cadmium telluride (CdTe), cupper-indium-gallium-selenide (CIGS), and the most common technology polycrystalline silicon (p-Si) and close to the performance of single crystalline silicon (s-Si) and silicon heterojunction (SHJ) [3]. Besides PCE, multiple techniques have been developed to produce the cells using liquid solutions to deposit the absorbing materials or gas phase deposition, allowing the possibility of having devices at lower costs [4, 5].

Perovskite refers to the material used in the absorber of photovoltaic cells. These are materials with the chemical formula ABX_3, named after the Russian mineralogist Lev Perov. For photovoltaic purposes A is usually an organic cation, B is commonly lead and X is a halide anion. Perovskites have been produced in research laboratories in configurations denominated single junction or in combination with other absorbing materials, to produce what has been called tandems. In the first case, the efficiency is constrained to around 33% by the Shockley-Queisser limit , while in the latter case the combination offers the possibility of increasing the efficiency beyond 40% [6].

Although one of the advantages of perovskite technologies has been the use of abundant materials in their absorber, a complete photovoltaic device requires additional layers with specific functions, such as collecting and conducting the charge carriers produced in the absorber or carrying the charges to an external circuit, in which they can be used. In the case of PSC, organic layers have been used to perform these tasks, but due to cost and stability issues these layers have been gradually changed to inorganic layers containing elements such as nickel or copper, among others [7, 8]. Additionally, transparent electrodes are also part of the cells [9]. Indium tin oxide and fluorine doped tin oxide have been traditionally used for this purpose [10].

The use of lead has been appointed as critical, due to the high toxicity of this element [11]. Other elements used for other layers are produced in limited amounts, leading to the motivation to quantify the amount of materials required for the future deployment of this technology and the amount of materials that can be dispersed into the environment because of this deployment. This work approaches the question of the material use and dissipation under different future scenarios, considering the future adoption of perovskite and the possibility of recycling of these materials. This work is not be misunderstood as a prediction but gives an overview of how the demand and the flows in the different life stages could develop if some of the assumptions made in this study are fulfilled in reality. Although the assumptions made for the study are as realistic as possible, high uncertainties remain in all the life stages in all scenarios.

2 Methodology

A material flow analysis (MFA) was done using the free "subSTance flow Analysis" (STAN) software developed by TU Vieanna that performs this kind of analysis according to the Austrian standard ÖNORM S 2096 [12]. The

A. Pehlken et al. (Eds.), *Cascade Use in Technologies 2018*,
https://doi.org/10.1007/978-3-662-57886-5_15

commercialisation of modules containing PSC technologies is forecasted to start in 2020. A horizon of 30 years was considered, since the life time of current commercial technologies is between 25 and 30 years. Hence the calculations were done for a horizon between 2020 and 2050. In the following sections, the different assumptions and methods used will be explained.

2.1 Materials Analysed

Two metals contained in two different layers were analysed in this work. The first one was the lead contained in the perovskite absorber, because this metal is one of the main components of the absorber layer. Although this material is not considered critical, it is considered hazardous and its use is restricted by the European Union in electronic components according to the Directive 2011/65/EU [13]. The stoichiometry employed in the calculations as an approximation to the formulation currently used in research was $FA_{0.85}MA_{0.15}PbI_3$, where FA is formamidinium, MA is methyl ammonium, Pb is lead and I is iodine [14]. Absorbers containing Caesium and Rubidium, instead of organic cations, have exhibited a higher stability regarding, but these materials are rather scarce, and quantities of annual production are not available [15]. The second material was Indium, which is mainly employed in the highly-transparent conducting material indium tin oxide (ITO) and is commonly deposited on glass as conductive layer or, in the case of monolithic tandem cells, as contact between the perovskite and the lower cell [16][17]. ITO is usually prepared from a mixture of Indium and Tin typically with a composition of 10:1. This material is considered as a critical raw material according to the European Commission [18]. Table 1 summarizes the annual production of the materials, when available, and the reserves of each material according to USGS.

Table 1. Annual production and reserves of different materials used in perovskite solar cells in 2017, taken from USGS [15]

Material	Production Tonnes/year	Reserves Tonnes
Lead	4.700×10^6	88.000×10^6
Iodine	31.000	6.400.000
Indium	720	-
Caesium	-	90.000
Rubidium	-	90.000

2.2 Market Scenario and Installation of Modules per Year

The installation of photovoltaic modules was assessed from the perspective of a 100% renewable energy scenario. According to the Lappeenranta University of Technology energy system model, an installed photovoltaic capacity of 11.958 GWp is expected for 2030, 27.400 GWp for 2050 and 42.000 GWp for 2100 [19]. The points between these values were obtained using the "smoothspline" algorithm contained in the software MATLAB ®. The installation rate per year was further estimated employing equation 1.

$$C_t = P_t - P_{t-1} \qquad (1)$$

Where,
C_t is the installation rate per year in the year t,
P_t is the total installed capacity in the year t,
P_{t-1} is the total installed capacity in the year t-1

The total installed capacity and the installation rate obtained from this method are shown in Figure 1.

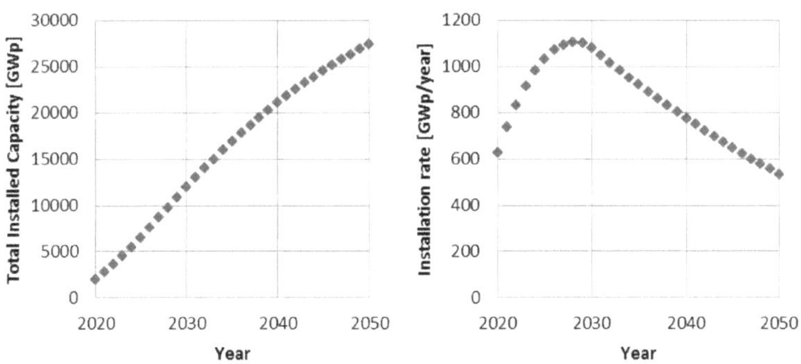

Figure 1. Total photovoltaic installed capacity and annual installation rate according to the interpolation and data from Breyer et al. [19]

For the purpose of this work, two levels of market penetration where considered for perovskite technologies, increasing linearly from 1% in 2020 to 20% and 80% in 2050, as a way to cover a scenario of low and high market penetration.

2.3 Material Flow System and Transfer Coefficients

Previous studies described the material flow system for CdTe and CIGS technologies [20, 21]. Although the production of PSC might differ in the type of deposition processes used to produce absorber layers, the material flow system would have similarities with the ones previously studied for other technologies, because PSC is also a thin film technology. Hence a similar system was used for the calculations done in this work and is summarized in Figure 2.

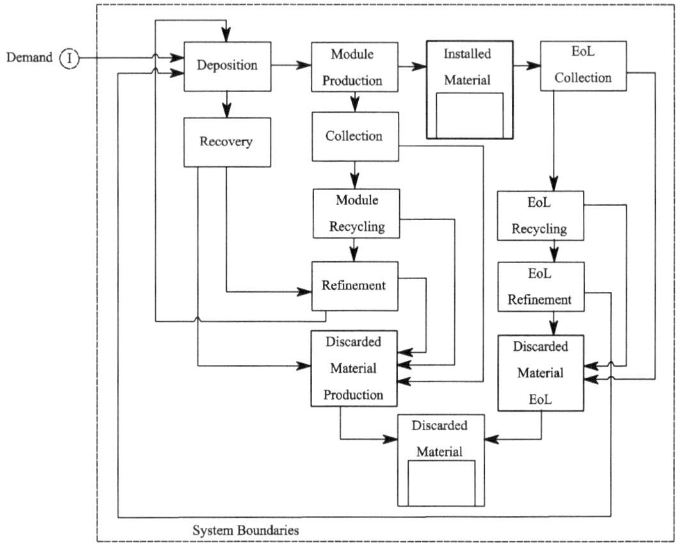

Figure 2. Material Flow System for Perovskite Solar Cells, adapted from Marwede et al. [21]

As depicted in Figure 2, the materials are initially used to perform the deposition process. A part of the material is deposited to produce cells, while the other part is not deposited and goes to a recovery phase. From this stream, a part is refined, while the other is discarded. The deposited material is later used to produce modules that are the final product. A part of these modules is defective and is further collected, recycled and the materials contained are refined to be reused in the deposition process, while another part goes directly to the discarded material. The use phase is represented by the installed material. In this process, the material is accumulated until the modules reach the end of life. After reaching the end of life, the material is collected, a part is recycled, and a part is directly discarded. In the subsequent processes, the material is recycled and refined to be again used in the deposition process or discarded. Finally, all the discarded material is accumulated in a virtual step called discarded material, although in a real situation this would be dissipated to the environment.

The transfer coefficients describe the proportion of the total input of material that is directed towards one output stream. As depicted in Figure 2, 11 processes redirect the materials to other processes, while two processes accumulate materials. The processes named "Discarded Material Production" and "Discarded Material EoL" gather inputs form other processes but have only one output. Hence, the transfer coefficient for these two processes towards the discarded material is 1. The transfer coefficients for other processes are summarized in Table 2. Values for all the years were obtained with linear changes between the beginning and the end of the calculation. Most of the transfer coefficients were chosen considering the study previously done by Marwede et al [21].

Table 2. Annual production and reserves of different materials used in perovskite solar cells

Transfer Coefficient		Lead		Indium	
From	To	2020	2050	2020	2050
Deposition	Module Production	0,80	0,95	0,17	0,50
Recovery	Refinement	0,50	0,75	0,80	0,93
Module Production	Installed Material	0,95	0,97	0,95	0,97
EoL Collection	EoL Recycling	0,80	0,90	0,80	0,90
EoL Recycling	EoL Refining	0,85	0,90	0,85	0,90
EoL Refining	Deposition	0,90	0,90	0,90	0,90
Collection	Module Recycling	0,95	0,99	0,95	0,99
Module Recycling	Refinement	0,90	0,90	0,90	0,90
Refinement	Deposition	0,85	0,90	0,85	0,90

Slot dye coating was considered as the most suitable process to perform the deposition of the perovskite absorber due to the high use of precursors. According to one equipment manufacturer, the material use for this process is over 95% [22]. As a conservative approximation, 80% was considered for 2020 increasing to 95% in 2050. ITO is usually deposited using sputtering. Two technologies are usually employed to do this process: using static planar targets and rotary targets containing a mix of indium and tin. This target is eroded, and the atoms are transferred to the material to be coated. The performance of the process is measured by the target utilization and the substrate collection efficiency. The first aspect is the amount of material eroded from the target compared to the initial amount of material, whilst the second aspect is the amount of material from the target that coats the desired surface. Lippens and Muehlfeld describe a target utilization of 34% and a substrate collection efficiency of 50% for planar target technology, while an optimal target utilization of 87.5% and a collection of 57.5% is expected for rotary targets [16]. The initial condition in 2020 was assumed as using planar target technology. Hence a total of 17% of the material is finally deposited on the cell surfaces, 66% remains in the targets and the rest is deposited in other parts of the equipment. The targets were assumed as totally recoverable, while 80% of the material sputtered elsewhere was assumed to be recovered. Hence, the initial transfer coefficient from deposition to module production of indium was assumed at 17% and the one from recovery to refinement at 80%. The optimum condition with rotary targets was assumed for 2050, considering that 50% of the target is deposited on the cells, all the remaining material in the targets are recovered and 90% of the material that is not deposited in the cells is recovered.

2.4 Material Embedded in Cells

The material embedded in the modules was calculated according to the stoichiometry of the different layers considering the descriptions made by different authors. The inflow of material was calculated employing equation 2, according to the approach of Marwede et al. [21].

$$F_{in}(t) = \frac{d_t \rho w_r}{I \eta_t} C_t \qquad (1)$$

Where,

F_{in} is the input of material
d_t is the thickness of material for the year t
ρ is the density of the material (4159 kg/m^3 and 7140 kg/m^3 for perovskite absorber and ITO respectively [23][24])
w_r is the mass concentration of the material in the layer, according to its stoichiometry
C_t is the installed capacity during the year t
I is the irradiance in standard testing conditions (1000W/m^2)
η_t is the efficiency of the module in the year t (16 – 24% for single junction, 25 – 30% for tandem)

Two architectures for perovskite solar cells were considered. The first architecture was in single junction, meaning that only one perovskite layer is used as absorber in the cell. In addition, tandem layers contain multiple layers that act as absorbers, allowing in theory a higher conversion of the incident light into electricity. However, more layers are included in the process to manufacture the cells. This can be seen in Table 3, where the tandem has more ITO layers compared to the single junction cells. To model improvements in the manufacturing of the cells, the thickness of the material and the efficiency of the module were varied linearly along the simulated years to simulate the optimization and improvement of the technology for the perovskite layer. In the case of the ITO layer, the thickness was considered constant through the calculations. The assumptions regarding these two parameters are summarized in Table 3 and were mainly taken from the architecture proposed by Bush et al for single junction perovskite solar cells and monolithic tandems containing a perovskite solar cell and a crystalline silicon cell [14].

Table 3. Thickness assumptions for different layers of single junction and tandem PSC/Si cells for the years 2020 and 2050

Layer	Single Junction		Tandem PSC/Si	
	2020 (nm)	2050 (nm)	2020 (nm)	2050 (nm)
Perovskite Absorber	500	300	500	300
Glass ITO contact	180	180	-	-
Bottom ITO contact	-	-	20	20
ITO contact between PSC and Si cell	-	-	20	20
ITO top contact	-	-	150	150

2.5 Life Time of the Modules

The life time of the modules was appointed using a Weibull in a cumulative distribution function, according to the modelling done by IRENA [25]. The proportion of modules that reach the end of life was calculated as the difference between the cumulative year for a year and the cumulative value for the previous year, as depicted in equation 3.

$$f_k(t) = \left(1 - e^{-((t-k)/T)^{\alpha}}\right) - \left(1 - e^{-(((t-1)-k)/T)^{\alpha}}\right) = e^{-(((t-1)-k)/T)^{\alpha}} - e^{-((t-k)/T)^{\alpha}} \quad (3)$$

Where,
$f_k(t)$ is the proportion of modules installed in the year k that reach the end of life in the year t
α is the shape factor of the Weibull distribution (α=5.3759 [25])
T is the average lifetime (Lifetimes of 5 years and 30 years was considered in this study)
t is the year in which the calculation is done
The total output of material was calculated using equation 4.

$$F_{EoL}(t) = \sum_{k=2020}^{k=t} f_k(t)\, F_{in}(k) \quad (4)$$

Where,
$F_{EoL}(t)$ is the total output of material in the year t
$f_k(t)$ is the proportion of modules installed in the year k that reach the end of life in the year t
$F_{in}(k)$ is the input of material in the year k

2.6 Scenarios

Four aspects were explored to create the different scenarios: perovskite technologies, market penetration of perovskite technology in single or in tandem configuration, lifetime of the modules and finally disposal at the end of life with recycling or without it. As to PSC, two configurations were considered: single junction and tandem of perovskite topping silicon solar cells. In the case of single junction an initial efficiency in 2020 of 16% was assumed, while for 2050 the efficiency assumed was 24%. In the case of the tandem PSC/Si, 25% and 30% were assumed as initial and final efficiencies respectively. The market penetration was considered in two levels: 20% and 80%. The mean lifetime was considered 5 or 30 years, respectively, and finally the cases considering no recycling at the end of life were done making the transfer coefficient 0 to discard all the modules. Considering all these aspects, 16 scenarios were generated. The scenarios were codified using the codes summarized in Table 4 and in the order Technology – Market Penetration – Lifetime of modules – Disposal at the end of life.

Table 4. Scenario codes description

Aspect of the scenario	Code description
Technology	SJ: Single Junction T: Tandem PSC/Si
Market penetration in 2050	20: market penetration from 1% to 20% 80: market penetration from 1% to 80%
Lifetime of the modules	R: regular end-of-life with T=30 years A: accelerated end-of-life with T=5 years
Disposal at the end of life	RC: Recycling NR: No recycling

2.7 Limitations of the Method

No replacement of modules after the end-of-life was considered. This can lead to errors, since the modules used for replacements also require materials to be produced. However, it is uncertain if the modules will be replaced by the same technology or a different one. Another assumption was that no material is lost during the use phase of the module due to failure of the encapsulation, which can lead to the dissipation of materials into the environment.

3 Results

3.1 Technology Impact

Figure 3 shows the requirements of lead necessary for the two perovskite technologies considered in this work in scenarios of adoption of 20% and 80% by 2050. The demand for lead is always lower for the tandem technology, due to the higher achievable efficiency. In all cases the demand reaches a maximum between 2033 and 2035, due to the decrease in the installation rate and the impact of recycling as a new supply. The demand for lead in the highest case reaches 1.094 tonnes, representing approximately 0,02% of the total annual production in 2017. Hence, the impact of the demand for lead for photovoltaic purposes might not have an important impact on the supply of this material. If the highest adoption scenario was considered for the single junction case, there would be a demand of 24.735 tonnes of lead for an installation of 9.645GW$_p$ of modules, which would represent approximately 2 days of current lead production. These results are in line with the predictions close to one day previously made by Jean et al [26]. On the other hand, the lead contained in discarded material might reach an amount of 2.229 tonnes in 2050, representing approximately 9% of the total lead demanded by the system in the 30 years of calculation. This amount can be reduced through collection after the use, improvement in the refining and recycling processes and improvements in the production processes.

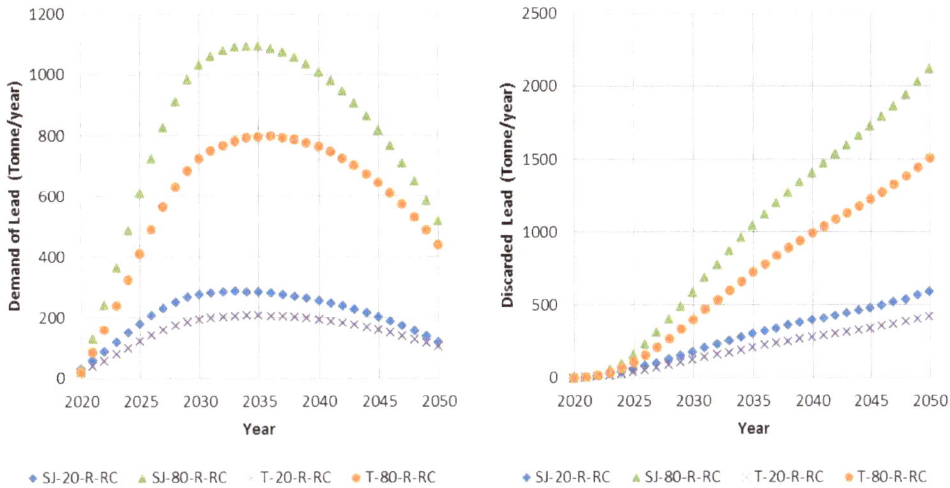

Figure 3. Demand for lead (left) and discarded lead (right) as a function of module technology and adoption in a scenario of normal lifetime and recycling at the end of life

Figure 4 depicts the demand for Indium for both technologies under the two adoption scenarios. In both cases, the single junction modules require a higher demand for Indium compared to the tandem modules. As in the last case, this is mainly due to the higher efficiency assumed for the tandem modules, making the used material per unit of peak capacity lower. In the 20% adoption scenario, the demand of this material reaches a maximum demand of 704 tonnes per year, close to the current estimated supply of 720 tonnes per year, as indicated in Figure 5 with a black line. With a higher adoption, the requirement of this material exceeds with up to 272% the current production of Indium. Although not presented here, the cumulative discarded indium in 2050, for the higher adoption case, single junction technology, regular lifetime of 30 years and recycling at the end of life amounts to 21.708 tonnes.

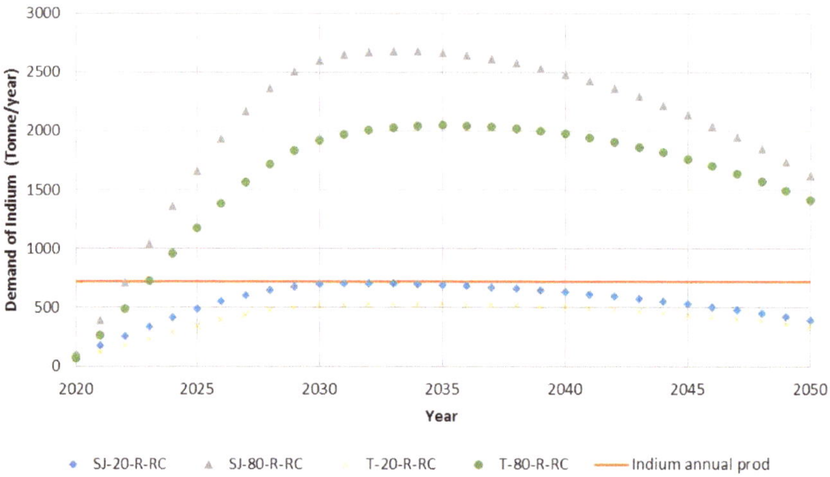

Figure 4. Demand for indium as a function of module technology and adoption in a scenario of normal lifetime and recycling at the end of life in comparison to the annual production in 2017

3.2 Lifetime impact

Lifetime of modules can also have an important impact on the demand for materials, particularly if they are recycled and incorporated again in the production. Figure 5 shows this effect for scenarios of 5 and 30 years of lifetime. In the case of a higher lifetime, the demand is higher because less material becomes available form old modules to be reprocessed and transformed into new modules. Yet, more material is also discarded due to the prompt reprocessing of these modules. For instance, the discarded material for a scenario of reaching a market share of 80% in the year 2050 with a lifetime of 30 years equals the discarded material for the scenario with a market share of 20% for the year 2050 and 5 years of lifetime. However, this work does not consider the replacement of modules, which would require more material and would generate more waste.

A different trend depicted in Figure 5 (left) is the anticipated peak of demand in the case of the scenarios considering a short lifetime. While the maximum demand for lead in these cases is reached in 2028, in the case of long lifetimes it is reached between 2033 and 2034, due to the anticipated supply of lead from recycled modules.

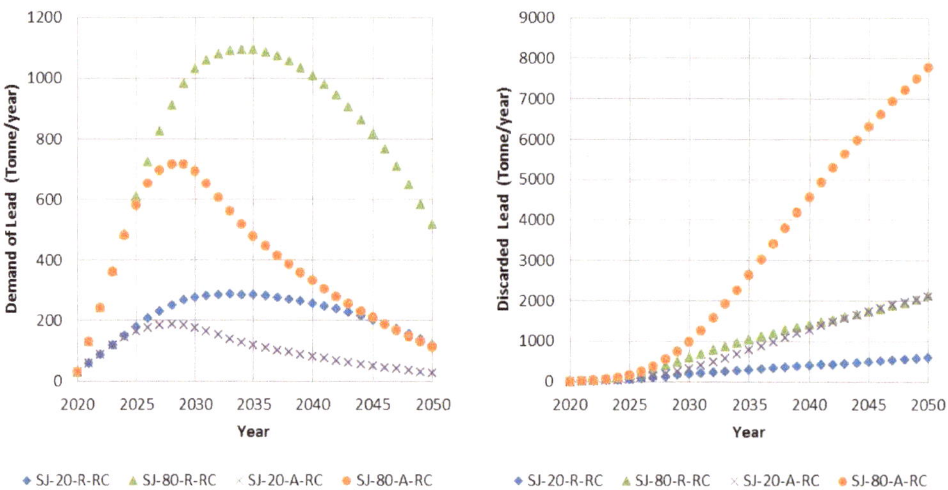

Figure 5. Demand for lead (left) and discarded lead (right) for single junction modules for low and high adoption scenarios under consideration of a lifetime of 30 years and 5 years

Figure 6 shows the results regarding indium, showing a similar qualitative result when compared to the results obtained for lead. However, both demand and discarded material are higher amounts in this case, exceeding again the current supply of indium in the cases with a higher market share. The discarded indium depicted in Figure 6 (right) has quantities an order of magnitude higher than those obtained for lead. This is derived from the higher losses in the deposition process, as summarized in Table 5. In this table, lead and indium are compared, showing that in the cases with a lifetime of 30 years, most of the material is discarded from the production process. When the lifetime is reduced, a lower amount of material is discarded from the production process compared to the material generated at the end-of-life.

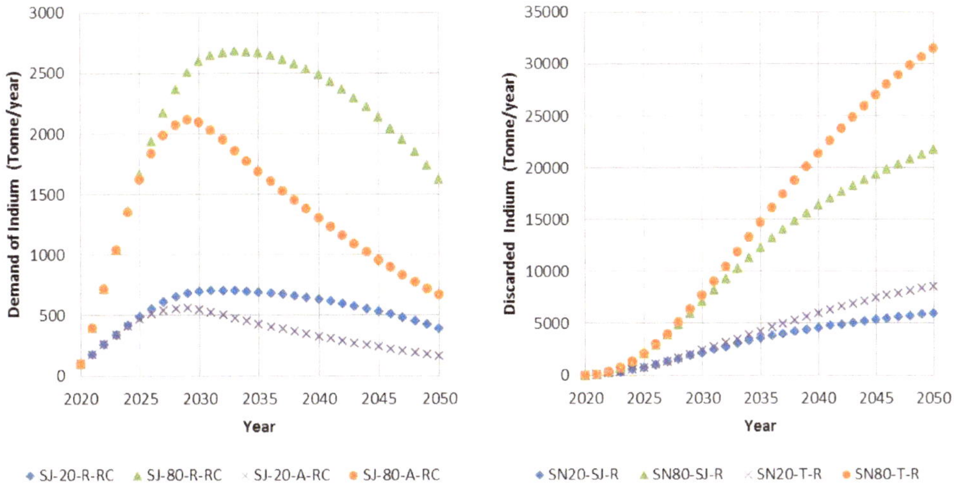

Figure 6. Demand for indium (left) and discarded indium (right) for single junction modules for low and high adoption scenarios under consideration of a lifetime of 30 years and 5 years

Table 5. Cumulative material discarded from production in comparison to the total discarded material in 2050

Material	Lead				Indium			
Scenario	SN20-SJ-R	SN80-SJ-R	SA20-T-R	SA80-T-R	SN20-SJ-R	SN80-SJ-R	SA20-T-R	SA80-T-R
Cum. material discarded from prod. (%)	81,21	83,03	22,88	22,65	98,29	98,7	68,44	68,02

3.3 Impact of Recycling at the End-of-Life

Figure 7 shows the impact of recycling on the demand for lead and the amount of lead discarded, which is particularly pronounced once the modules start to reach the end of life. For the case with low penetration of perovskite technology, the discarded lead increases to 48%, whilst in the case of high market penetration the increase is 43%. Figure 8 shows the lower impact of recycling in the case of indium, generating an increase of approximately 7% in the discarded quantity in both market penetration cases. This low sensitivity in the case of indium derives again from the high losses in the manufacturing process, compared to the loss of material at the end of life.

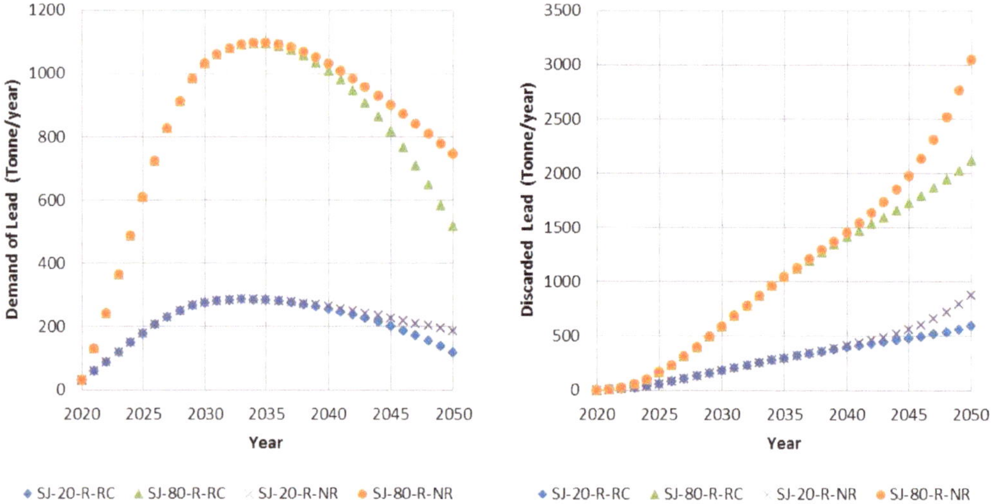

Figure 7. Demand for lead (left) and discarded lead (right) for single junction modules for low and high adoption scenarios with and without recycling

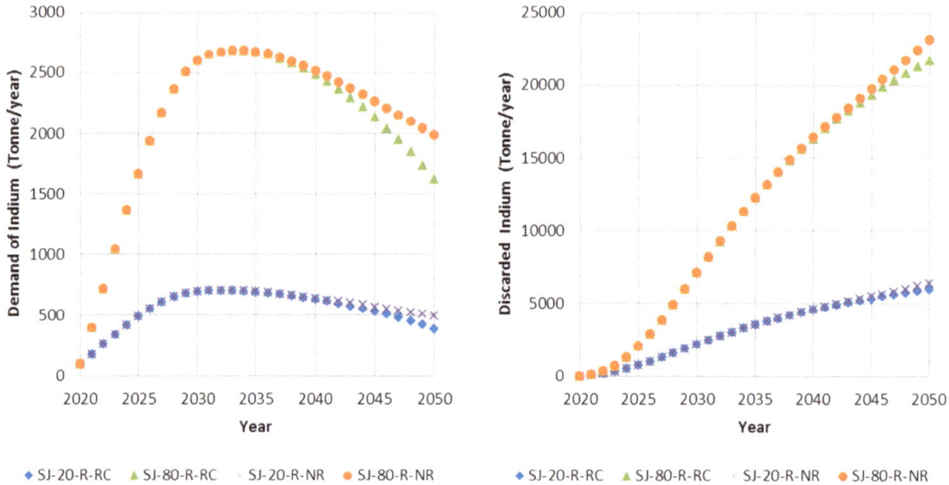

Figure 8. Demand for indium (left) and discarded indium (right) for single junction modules for low and high adoption scenarios with and without recycling

3.4 Summary of all Scenarios

Table 6 compiles the maximum demand and the discarded quantities for both materials and all the scenarios. The maximum demand for lead for single junction cases was 1.096 tonnes, whilst for the tandem it was 799 tonnes, representing in that case a reduction of 27%. In the case of indium, the maximum demand calculated was 2.684 tonnes for the single junction cell, while for the tandem it was 2.052 tonnes, representing a reduction of approximately 24%. Nevertheless, the most important variations between the scenarios are in the discarded amount of materials. For instance, in the scenario for single junction, high market penetration, long lifetime and no recycling at the end of life, the discarded lead is 9.9 times the amount discarded in the scenario for single junction, high market penetration, long lifetime and recycling. Hence, in case of adoption of this technology, the recycling must play an important role to avoid high environmental effects from the dispersion of lead into the environment.

Table 6. Summary of maximum demand and discarded material for all the scenarios studied in this work

Scenario	Max. demand Pb (t/y)	Discarded Pb 2050 (t)	Max. demand In (t/y)	Discarded In 2050 (t)
SJ-20-R-RC	285	591	704	5953
SJ-80-R-RC	1.094	2.119	2.679	21.708
SJ-20-A-RC	186	2.098	558	8550
SJ-80-A-RC	715	7.770	2.110	31.499
SJ-20-R-NR	277	876	705	6.385
SJ-80-R-NR	1.096	3.052	2.683	23.134
SJ-20-A-NR	286	5.626	705	14.645
SJ-80-A-NR	1.096	20.974	2.684	54.455
T-20-R-RC	206	418	531	4.483
T-80-R-RC	796	1.506	2.048	16.430
T-20-A-RC	130	1.503	413	6.467
T-80-A-RC	503	5.588	1.558	23.928
T-20-R-NR	207	610	533	4.793
T-80-R-NR	799	2.140	2.052	17.457
T-20-A-NR	207	4.048	533	11.129
T-80-A-NR	799	15.139	2052	41.532

4 Conclusions

The main aim of this work was appraising the amount of lead and indium required in the future for the commercialization of perovskite solar cells. This work proved that under current conditions, perovskite solar cells are not constrained by the supply of lead used in the absorber. On the other hand, the use of indium tin oxide for the transparent conducting layers may increase the demand for indium considerably beyond current production. Since indium is not mined directly and is recovered from the production of zinc, these could cause a supply constraint. However, substitute materials might be used to reduce this risk. A maximum demand for lead of 1.096 tonnes and a maximum demand for indium of 2.684 tonnes were obtained as a result of the calculations done for modules containing single junction perovskite solar cells, a market share of 80%, mean lifetime of 5 years and no recycling. Most of the variation was presented in the discarded material, showing that recycling and lifetime extension can minimize the quantity of materials that is dissipated into the environment and can have potential environmental impacts.

This work also proved that the use of new architectures, such as tandems of perovskite absorbers with silicon, can decrease the quantity of lead and indium that is used per GW_p of modules, due to the higher efficiency of the cells. Although this reduction is desirable, the additional processing stages and the energy use in the manufacturing processes can impact the costs and environmental impact of these novel cells and in the long run their adoption as main photovoltaic technology.

Increasing the lifetime from 5 to 30 years may in the long run increase the demand for lead and indium due to the absence of recycled material to supply the production of new modules, but can also prevent the dissipation of these metals into the environment and their negative effects.

The discarding of indium was found to be more dependent on the production process, since the transfer coefficients for the deposition are rather low due to multiple factors related to the sputtering process. Hence, the recycling at the end of life has less impact on the demand of material and discarding of material, compared to the case of lead. However, this is a motivation to either replace this material, increase the efficiency of deposition or decrease its use in the cells via thickness reduction of the layers.

These kinds of calculations contain many assumptions and uncertainties. Although they should not be misunderstood as a prediction of the future, they can provide an idea of the effect of different aspects in the material requirements. The development of perovskite solar cells is a dynamic field of research. Hence, with new architectures and materials, these calculations must be updated to deliver realistic results in the future.

5 Zusammenfassung

In diesem Beitrag werden mittels einer Materialflussanalyse Bedarf und Ausschuss von Blei und Indium in Szenarien der zukünftigen Einführung von Perowskit-Solarzellen berechnet. Für die Erstellung der Szenarien wurden vier Faktoren berücksichtigt: Zum einen wurden als Typen der Perowskit-Solarzelle entweder Einzelverbindungszellen oder Tandems mit Silikonsolarzelle angenommen. Der angenommene zukünftige Marktanteil lag zwischen 20% und 80% in 2050. Als Lebensdauer der Module wurde entweder 5 Jahre oder 30 Jahre angenommen. Zuletzt wurden Szenarien mit und ohne Recycling evaluiert. Die Ergebnisse zeigen, dass der Bedarf an Blei voraussichtlich nicht signifikant im Vergleich zum bestehenden Marktangebot sein wird. Die Verwendung von Indium im Szenario mit hohem Marktanteil könnte dagegen das heutige Angebot übersteigen. Die Verwendung von Tandemtechnologien könnte den Materialbedarf aufgrund der höheren Effizienz in der Energieumwandlung verringern. Zudem könnte eine höhere Lebensdauer sowie ein Recycling am Ende des Lebenszyklusses die Menge an Material verringern, die an die Umwelt abgegeben wird.

6 References

[1] IPCC, "Climate Change 2014: Synthesis Report. Contribution of Working Groups I, II and III to the Fifth Assessment Report of the Intergovernmental Panel on Climate Change," Climate Change 2014: Synthesis Report. Contribution of Working Groups I, II and III to the Fifth Assessment Report of the Intergovernmental Panel on Climate Change, 2014. [Online]. Available: http://ipcc.ch/pdf/assessment-report/ar5/syr/SYR_AR5_FINAL_full_wcover.pdf. [Accessed: 18-Dec-2017].

[2] N. Jungbluth and M. Stucki, "Life cycle inventories of photovoltaics," ESU-services Ltd., 2012. [Online]. Available: http://www.esu-services.ch/fileadmin/download/publicLCI/jungbluth-2012-LCI-Photovoltaics.pdf. [Accessed: 12-Jan-2018].

[3] M. A. Green, Y. Hishikawa, E. D. Dunlop, D. H. Levi, J. Hohl-Ebinger, and A. W. Y. Ho-Baillie, "Solar cell efficiency tables (version 51)," Prog. Photovolt. Res. Appl., vol. 2018, no. 26, p. 20, 2017.

[4] I. Mesquita, L. Andrade, and A. Mendes, "Perovskite solar cells: Materials, configurations and stability," Renew. Sustain. Energy Rev., no. May, 2017.

[5] L. Qiu, L. K. Ono, and Y. Qi, "Advances and challenges to the commercialization of organic–inorganic halide perovskite solar cell technology," Mater. Today Energy, 2017.

[6] Q. Wali, N. K. Elumalai, Y. Iqbal, A. Uddin, and R. Jose, "Tandem perovskite solar cells," Renew. Sustain. Energy Rev., vol. 84, no. December 2017, pp. 89–110, 2018.

[7] W. Chen et al., "Efficient and stable large-area perovskite solar cells with inorganic charge extraction layers Efficient and stable large-area perovskite solar cells with inorganic charge extraction layers," vol. 350, no. November, pp. 1–6, 2015.

[8] R. Rajeswari, M. Mrinalini, S. Prasanthkumar, and L. Giribabu, "Emerging of Inorganic Hole Transporting Materials For Perovskite Solar Cells," Chem. Rec., vol. 17, no. 7, pp. 681–699, 2017.

[9] M. Anaya, G. Lozano, M. E. Calvo, and H. Míguez, "ABX3Perovskites for Tandem Solar Cells," Joule, vol. 1, no. 4, pp. 769–793, 2017.

[10] M. I. Asghar, J. Zhang, H. Wang, and P. D. Lund, "Device stability of perovskite solar cells – A review," Renew. Sustain. Energy Rev., vol. 77, no. April, pp. 131–146, 2017.

[11] A. Abate, "Perovskite Solar Cells Go Lead Free," Joule, vol. 1, no. 4, pp. 659–664, 2017.

[12] O. Cencic and H. Rechberger, "Material flow analysis with software STAN," J. Environ. Eng. Manag., vol. 18, no. 1, pp. 3–7, 2008.

[13] European Commission, "DIRECTIVE 2011/65/EU OF THE EUROPEAN PARLIAMENT AND OF THE COUNCIL of 8 June 2011 - ROHS," Official Journal of the European Union, 2011. [Online]. Available: http://eur-lex.europa.eu/LexUriServ/LexUriServ.do?uri=OJ:L:2011:174:0088:0110:EN:PDF.

[14] S. Yang, W. Fu, Z. Zhang, H. Chen, and C.-Z. Li, "Recent advances in perovskite solar cells: efficiency, stability and lead-free perovskite," J. Mater. Chem. A, vol. 5, no. 23, pp. 11462–11482, 2017.

[15] USGS, "Mineral Commodity Summaries 2018," 2018. [Online]. Available: https://minerals.usgs.gov/minerals/pubs/mcs/2018/mcs2018.pdf.

[16] P. Lippens and U. Muehlfeld, "Indium Tin Oxide (ITO): Sputter Deposition Processes," in Handbook of Visual Display Technology, J. Chen, W. Cranton, and M. Fihn, Eds. Cham: Springer International Publishing, 2016, pp. 1215–1234.

[17] K. a. Bush et al., "23.6%-Efficient Monolithic Perovskite/Silicon Tandem Solar Cells With Improved Stability," Nat. Energy, vol. 2, no. 4, pp. 1–7, 2017.

[18] European Commission, Study on the review of the list of critical raw materials, no. June. 2017.

[19] C. Breyer et al., "On the role of solar photovoltaics in global energy transition scenarios," Prog. Photovoltaics Res. Appl., vol. 25, no. 8, pp. 727–745, Aug. 2017.

[20] M. Marwede and A. Reller, "Estimation of Life Cycle Material Costs of Cadmium Telluride – and Copper Indium Gallium Diselenide – Photovoltaic Absorber Materials based on Life Cycle Material Flows," J. Ind. Ecol., vol. 18, no. 2, pp. 254–267, 2014.

[21] M. Marwede and A. Reller, "Future recycling flows of tellurium from cadmium telluride photovoltaic waste," Resour. Conserv. Recycl., vol. 69, pp. 35–49, 2012.

[22] MBraun, "Slot Dye Coating," 2018. [Online]. Available: http://mbraun.de/products/coating-equipment/sloit-die-coater/#specifications. [Accessed: 20-Feb-2018].

[23] C. C. Stoumpos, C. D. Malliakas, and M. G. Kanatzidis, "Semiconducting tin and lead iodide perovskites with organic cations: Phase transitions, high mobilities, and near-infrared photoluminescent properties," Inorg. Chem., vol. 52, no. 15, pp. 9019–9038, 2013.

[24] AZoM, "Indium Tin Oxide (ITO) - Properties and Applications," 2004. [Online]. Available: https://www.azom.com/article.aspx?ArticleID=2349. [Accessed: 23-Jun-2018].

[25] IRENA, "END-OF-LIFE MANAGEMENT /Solar Photovoltaic Panels," 2016. [Online]. Available: www.irena.org/menu/index.aspx?mnu=Subcat&PriMenuID=36&CatID=141&SubcatID=2734.

[26] J. Jean, P. R. Brown, R. L. Jaffe, T. Buonassisi, and V. Bulovic, "Pathways for solar photovoltaics," Energy Environ. Sci., vol. 8, no. 4, pp. 1200–1219, 2015.

Comparison of Cascaded Utilization with Life Cycle Assessment – a Case Study of Wind Turbine Blades

Kalle Wulf[1], Frauke Germer[1] and Henning Albers[1]
[1]Department of Civil and Environmental Engineering, University of Applied Sciences Bremen, Bremen, 28199, Germany
frauke.germer@hs-bremen.de

Abstract

The utilization of resources in multiple cascade stages can improve the resource efficiency of product systems. In order to compare one cascaded utilization with another, an assessment of the contribution of the cascaded utilization by quantifiable indicators is necessary. A product system with cascaded utilization has multiple utilizations in addition to the primary utilization. This paper discusses a method by which a comparison of cascaded utilizations can be achieved despite the multi-functionality of such systems. Based on a comparison of the cascaded utilization with an equivalence system, potential reductions in impact indicators can be identified.
In a case study, four scenarios for the cascaded utilization of a rotor blade are developed on the basis of a literature review. Potential reductions in the impact indicators Cumulative Energy Demand (CED) and Cumulative Raw Material Demand (CRD) are calculated. Significant reductions in primary energy demand are observed for the usage of recycled glass-fibres in secondary products as well as the usage of rotor blade material as secondary fuel in the cement industry. The greatest reduction in raw material demand resulted from the mechanical recycling of rotor blades and utilization of the material as a filler.
The main results of a review of the method are twofold: (i) The selection of the equivalence systems has a decisive influence on the result of the study and (ii) the approach requires a high data diversity and availability.

1 Introduction

The principle of cascaded utilization receives increasing interest in discussions about resource efficiency. The productivity of resource usage can be improved by multiple steps of material utilization followed by a final energy recovery. The *German Resource Efficiency Program* (ProgRess II) emphasizes this approach as a key factor for a sustainable use of resources. Primary resources can be substituted by the usage of secondary resources and fuels. Consequently, the environmental impacts in processes can be significantly reduced. However, an assessment of a cascaded utilization based on measurable indicators still needs to be developed [1].
The method of Life-Cycle Assessments (LCA) is used in order to assess environmental impacts of product systems. The exact definition of an LCA is specified within the standards ISO EN 14040 as well as 14044. Environmental impacts are assessed across all stages of a product's life cycle (*"cradle-to-grave"*), including the raw material extraction, manufacturing, use and disposal [2].
Within LCAs, product systems with cascaded utilization need to be considered as *multi-functional systems*. In cascaded utilizations materials stay within the product system after the stage of disposal and generate additional utilizations [3]. According to the definition of LCAs product systems are required to fulfil the same level of utilization to be comparable. Usually this is achieved by definition of a functional unit. However, a product system with cascaded utilization has multiple functional units. Thus, in order to compare one cascaded utilization with another, the method of LCA needs to be adapted. A possible solution to this problem will be discussed in this paper.

1.1 State of the Art

The cascaded utilization of materials is especially applied in current research questions within the field of renewable materials. In this context, the cascaded utilization of wood has been researched in various articles [3], [4].
However, potential cascaded utilizations of materials can also be identified for other product systems. Paraskevas et al. [5] presented the recycling of aluminum in cascades. Various scenarios for the cascaded utilization of paper are discussed in Korhonen et al. [6]. Höglmeier et al. [7] presented an approach for cascaded utilizations in life cycle assessments. A comparison of cascaded utilization of wood waste with the usage of new wood was considered. A sequential use of wood waste in the form of chipboard of different qualities with subsequent energetic use was compared with the production of chipboard of the same quality from new wood with direct energetic utilization. In order to achieve an equality of benefit, the method of system expansion was used.
In the literature representations of cascaded utilizations do not follow a uniform system. Against this background, a model for the presentation of cascaded utilization of materials is to be introduced. Subsequently, a methodology for the implementation of cascaded utilization within a life cycle assessment is described in this work. In addition, the proposed methodology will be applied to a case study.

A. Pehlken et al. (Eds.), *Cascade Use in Technologies 2018*,
https://doi.org/10.1007/978-3-662-57886-5_16

2 Material and Methods

2.1 Cascade Chain Model

Figure 1 shows the use of a resource or product in cascade stages in a two-dimensional model (quality and time). In this way, Sirkin and Houten first described the cascaded utilization in 1994 in a very abstract way by comparing it to a river flowing over a sequence of plateaus. The water falls continually from the one level to the next, until its forces eventually come to equilibrium at the lowest level in the cascade. This concept is applied to the cascaded utilization of resources. The life of resources is extended by sequentially exploiting the potential in stages, until no further utilization under consideration of technical and economic limitations can be achieved.

In this model, resource quality is defined as an expression of the capacity to perform various tasks at various degrees of difficulty [8].

Due to the multitude of uses, such a system provides a plurality of additional utilizations in addition to the primary utilization (commonly represented by the functional unit in LCAs). An additional utilization could be the supply of a mass of another product. In principle, more complex additional utilization could be defined as well (e.g. fulfilment of a certain task such as packaging for a defined volume). An additional utilization can also be provided by the prolonged use of a product or a resource. Figure 2 shows a summary of the utilizations of a cascaded product system. In the first cascade stage, the primary utilization is represented. Utilization from subsequent cascade stages is defined by additional utilizations. Utilization that results from the final utilization of the resource - for example through energy recovery - must also be taken into account.

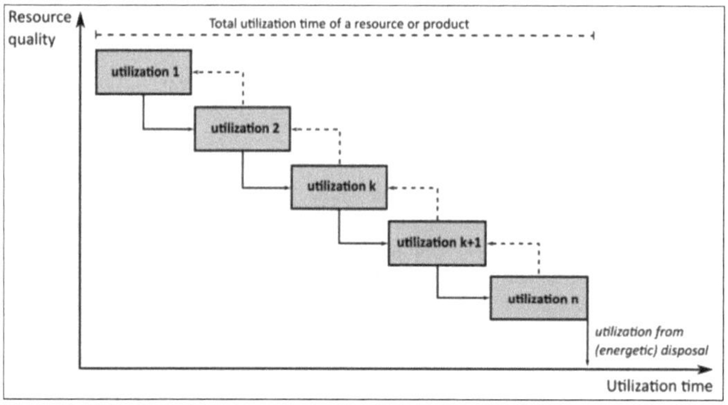

Figure 1. Utilization of a resource or product in cascade stages [8]

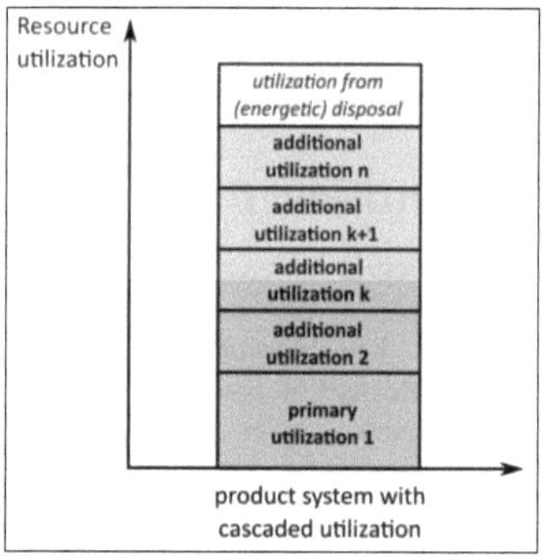

Figure 2. Total utilization of a product system with cascaded utilization

2.2 Comparison of Cascaded Utilization and Equivalence System

A product system with cascaded utilization is compared with an equivalence system, in which all utilizations are fulfilled by a plurality of individual product systems. This approach is shown schematically in Figure 3. A separate product system (System 1, System 2, etc.) is defined for each utilization provided by the cascaded utilization. These must be designed in such a way that they fulfil the same utilization as specified by the product system with cascaded utilization. If all utilizations are described by individual product systems, they can be defined together as an equivalent system to the product system with cascaded utilization.

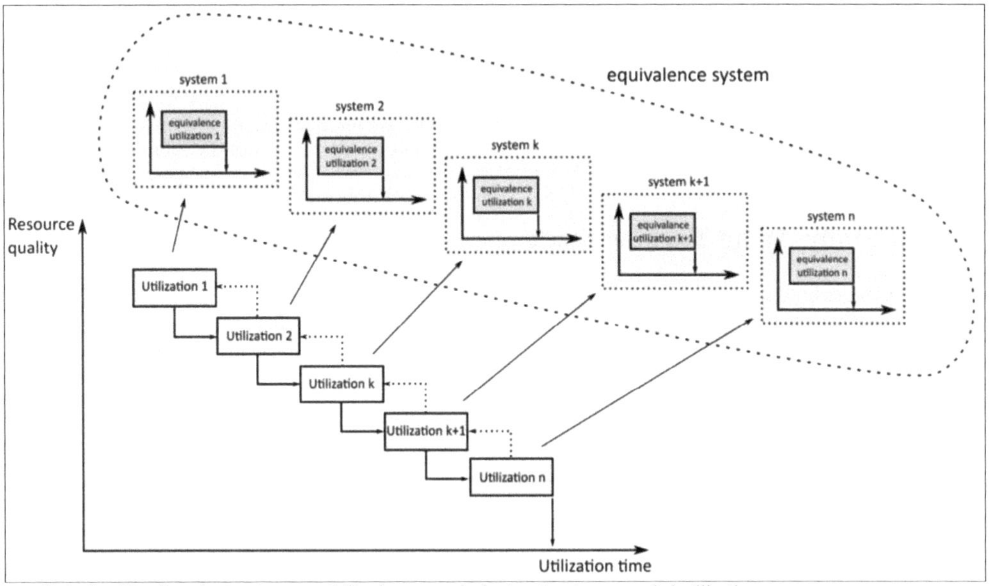

Figure 3. A plurality of product systems are defined as an equivalent system to a cascaded utilization.

2.3 System Expansion to Create Equality of Benefit

In order to enable a comparison between cascaded utilization and equivalence system, an equality of benefit of the systems needs to be established. This is done by system expansion. Figure 4 shows the utilizations of the equivalence system and the product system with cascaded utilization in comparison. In particular, a utilization resulting from a possible (energetic) disposal must be taken into account. Duo to the plurality of product systems this is usually higher in the equivalence system. For this reason, it is often necessary to further expand the system. Environmental impacts from a comparable product system must be assumed in this case.

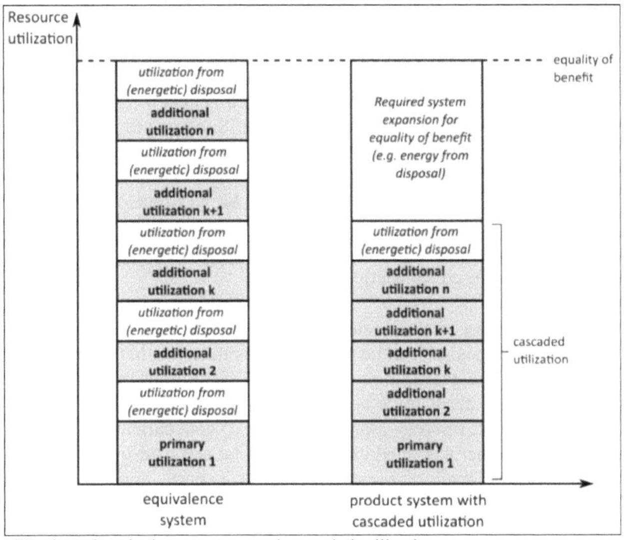

Figure 4. Comparison of utilization of equivalence system and cascaded utilization

2.4 Life Cycle Assessment

For both systems (product system with cascaded utilization and equivalence system), an impact assessment can be made as in common life cycle assessments. This seeks to "identify and assess the size and significance of potential environmental impacts of a product system in the course of its life cycle" [9], [10]. Environmental impacts are assessed by aggregating the outputs into impact categories, which are thus based on the inventory analysis of the study. The

selection of the impact categories considered is subject to the creator of the life cycle assessment. Impact categories are characterized by quantifiable impact indicators. These include in particular the cumulated energy demand (CED) and cumulated raw material demand (CRD) as indicators of the total primary energy demand and quantities of raw materials used in the product system, respectively.
Impact indicators of the product system with cascaded utilization and the equivalence system can be compared. Consequently, potential savings in impact categories by cascaded utilization can be identified. In this way, different scenarios for cascaded utilization are compared.

3 Case Study: Wind Turbine Blade

3.1 System under Study

The expansion of the wind energy sector is an essential pillar of the energy transition in Germany. The Renewable-Energy-Sources-Act (EEG) schedules the growth of renewable energies for the coming years [11]. An annual increase of 2,500 MW in the installed capacity of onshore wind energy is planned. For the offshore sector, 6,500 MW are planned in 2020 and 15,000 MW in 2030 (§3, EEG 2014).
The expansion of renewable energies involves a high material and energy demand for the production and maintenance of wind turbines. Furthermore, the resource and material efficiency of the product system is highly dependent on the recycling of the plants. While the foundation, tower, gearbox, generator and electronic components can largely be recycled via conventional disposal channels, there are only limited possibilities for the recycling of the rotor blades of the plant. Rotor blades are made of fibre-reinforced composites consisting of approximately equal parts of epoxy resin as well as glass and carbon fibres. At present, end-of-life turbine blades may be used in smaller mass flows to generate energy in waste-to-energy plants. A possibility for a combined energetic and material recycling is the use as secondary fuel in the cement industry [12]. However, recycling capacities are expected to be exceeded as early as 2019 [13]. The development or expansion of recycling capacities must therefore be regarded as necessary.
Against this background, scenarios for the cascaded utilization of a rotor blade of Enercon's gearless onshore wind turbine E-112 are examined. The rotor blade is manufactured in hand lay-up. The total mass of the rotor blade of approximately 20 tons is composed of glass fibre (41%), epoxy resin (38%), PU rigid foam (5%) and small amounts of aluminium (1%). Other materials (15%) are plastics such as polyamide, polyethylene and PVC as well as paint and rubber [14].

3.2 Scenarios for Cascaded Utilization

Scenarios for the cascaded utilization of rotor blades from wind turbines were developed based on a literature research. These include different types of recycling:

Scenario I: After utilization of the rotor blade to generate energy in a wind turbine, the material is reduced to smaller pieces for use as a filler. This step-by-step process consists of pre-treatment at the wind turbine site using mobile cutting technology and shredding using a hammer mill or cutting machine. Subsequently, the particle size is further reduced with the aid of a hammer or high-speed mill [12], [15]. Granules resulting from such a process can be used as a substitution product in the production of bulk molding compounds (BMC) and sheet molding compounds (SMC). In this way, up to 10% of calcium carbonate and glass fibres can be substituted with negligible impairment of the mechanical properties [16]. A further cascaded utilization could be achieved by repeated shredding/grinding of the BMC/SMC component and reuse as filler [17]. In this way, calcium carbonate can be substituted in the production of new BMC/SMC. Finally, the material is used for energy recovery.

Scenario II: In the pyrolysis process, organic substances are thermally decomposed in the absence of air (depolymerisation). The resulting pyrolysis gas can be used as a substitute fuel in the process. The pyrolysis oils can be used as a homogeneous feed material in a gasification process [18]. The fibrous materials (glass and carbon fibres) can be reused. However, a new silane layer needs to be applied to the chopped fibres. Recycled fibres can be used in the production of sheet molding compounds, whereby high proportions of conventional fibres can be replaced by recycled fibres [19], [20]. As in scenario I, another cascaded utilization can be assumed by shredding/grinding the BMC/SMC component and repeated usage as filler. Finally, the material is used for energy recovery.

Scenario III: In this scenario, fibres are recovered by pyrolysis and used as insulating material. This method has already been used by the company ReFiber Aps from Denmark. The recovered fibres were mixed with a small amount of polypropylene fibres and reheated. By melting the polypropylene fibres, the glass fibres combine to form a glass fibre wool. This can be used as heat-resistant insulation material or filler [12]. After use as glass wool, landfilling of the material is assumed.

Scenario IV: End-of-life rotor blades can be used as secondary fuel in the cement industry. Pulverized lignite, limestone and quartz sand can be substituted in this process. For this purpose, the rotor blade segments are shredded to a size of less than 50 mm. After separation of metals and other impurities, the material is mixed with a second material stream (e.g. rejects from the paper industry) for homogenization and dust reduction [21]. The product cement can be used for the construction of a new wind power plant foundation. Finally, the material is assumed to be deposited.

Figure 5 shows the described scenarios for cascaded utilization. Hereafter, these are to be compared within a life cycle assessment using the methodology previously specified.

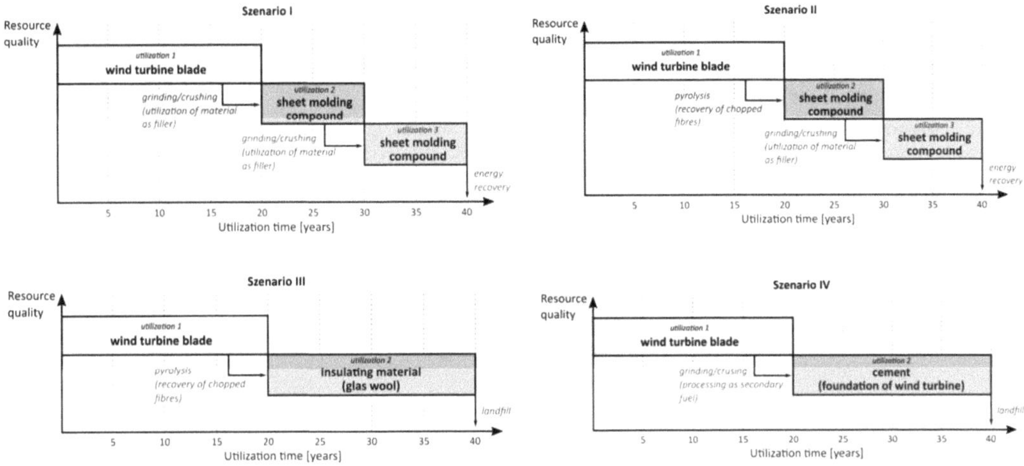

Figure 5. Scenarios for the cascaded utilization of a rotor blade (Note that resource quality is not quantified in this study, therefore the resource quality between scenarios cannot be compared.)

3.3 Data Sources

Impact indicators are calculated based on the black box method. Accordingly, material flows and processes that are identical both in the product system with cascaded utilization and the equivalence system have been neglected.Energy demand for shredding processes were estimated in accordance with the work of Friedel (1999), in which the author describes the primary energy demand in order to achieve certain pieces sizes depending on material types. The material type of plastics was assumed for the shredding/crushing of the rotor blades.

For transports, a distance of 150 km was assumed both to the recycling plant and to further manufacturing processes (e.g. SMC component). Primary energy consumption is calculated on the basis of basic data according to the mass to be transported [23].Pyrolysis processes for fibre recovery were modelled based on a Best Available Technique Reference document. A pyrolysis plant connected to a power plant for electricity generation was assumed. Thermal energy resulting from the combustion of pyrolysis oil and coke is used both in the pyrolysis process and in the power plant [24].Data on most of the materials (e.g. natural gas, hard coal, glass fibre, calcium carbonate, limestone, quartz sand) were taken from the database ProBas (process-oriented basic data for environmental management systems) of the German Federal Environment Agency (Umweltbundesamt). In some cases, data from comparable processes were used, unless specific information was available for the recycling of rotor blades. This applies in particular to shredding/crushing processes and the pyrolysis process.

3.4 Impact Categories

The indicators Cumulative Energy Demand (CED) and Cumulative Raw Material Demand (CRD) are selected as impact indicators for the combined assessment of material and energy demand of the product system.Cumulative energy demand indicates the total primary energy demand that results from or can be allocated to the production, use and disposal of an economic good (product or service) [25]. As a mass-related efficiency indicator, the Cumulative Raw Material Demand describes the sum of the quantities of raw materials required to supply an economic good. Similar to the CED, the raw material demand includes all stages of the life-cycle (production, use and disposal). An additional criticality analysis aims at the identification of raw materials with essential function for the product system. The vulnerability of a reference system (e.g. of a company) to supply disruptions of specific raw materials needs to be assessed and is represented by a weighting factor based on numerous criteria of criticality [26].

3.5 Results of the Case Study

Figure 6 shows reductions in cumulative energy demand by cascaded utilization of the rotor blade. The greatest reduction in primary energy demand is therefore found in Scenario IV through combined material/energy recycling of the material in the cement plant. This is primarily due to the substitution of pulverized lignite in the cement production process. Scenarios II and III also show potential reductions through the substitution of primary energy-intensive glass fibres in the production of glass wool or SMC components. When using the rotor blade material as filler in Scenario I, on the other hand, calcium carbonate and only small amounts of glass fibres can be replaced in the production of SMC components. Due to the comparably lower primary energy consumption to provide calcium carbonate, there is also a lower reduction in energy demand identified.

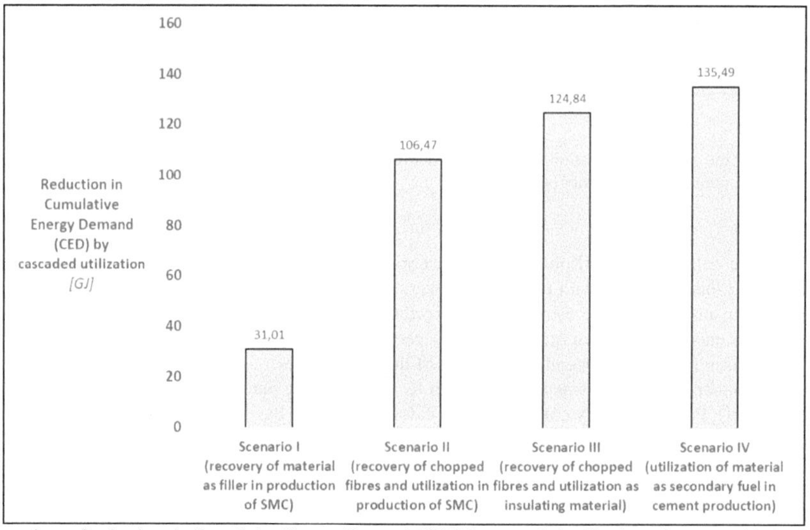

Figure 6. Comparison of reductions in Cumulative Energy Demand (CED) by cascaded utilization of one rotor blade of the wind turbine E-112

Figure 7 shows reductions in the cumulative raw material demand for the scenarios under consideration. In Scenario I the greatest reductions are identified, resulting from the substitution of calcium carbonate in the manufacture of SMC components. In scenarios II and III, the application of the pyrolysis process results in lower reductions in raw material demand, since only the recycled glass fibres are used in subsequent processes. In Scenario IV, too, the potential reductions for raw material demand are lower.

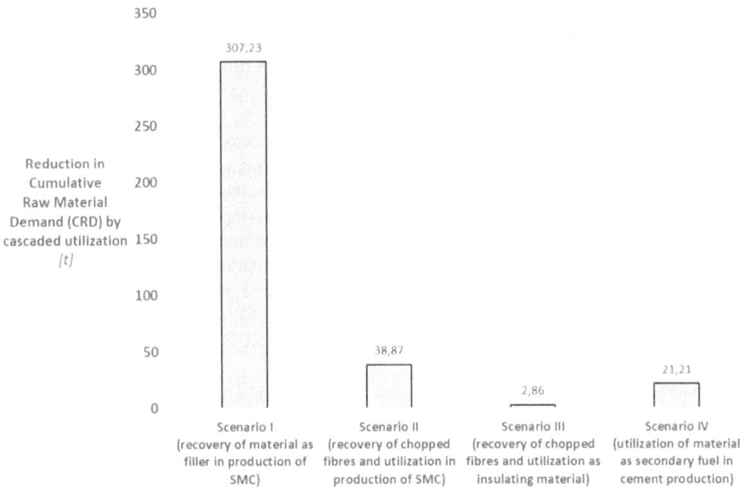

Figure 7. Comparison of reductions in Cumulative Raw Material Demand (CRD) by cascaded utilization of one rotor blade of the wind turbine E-112

4 Discussion

The proposed methodology will be discussed below. Experiences from the application of the method to the case study of the cascaded utilization of a rotor blade are provided. In this way, both the limits of the method and problems in its implementation are to be identified, so that, if necessary, further development of the method is possible:

(i) Selection of equivalence systems
The selection of the equivalence systems has a decisive influence on the result of the study. As described in section 2.2, all utilizations of the product system with cascaded utilization need to be represented by individual product systems. However, there is not always only one possible process, but products can be manufactured in different ways. Clear rules must be defined here (e.g. always choice of the process which corresponds to the state of the art). Nevertheless, this can lead to inaccuracies and needs to be observed closely. In addition, detailed data is often not available, which can lead to further uncertainties since comparable processes need to be assumed.

(ii) Achieving uniform timeframes
Achieving uniform timeframes for all scenarios with cascaded utilization can be problematic. However, in order to be able to compare scenarios with each other, all must extend over the same time period. In some cases, this requires assumptions that only partially reflect reality. For example, further cascade stages must be assumed to extend the utilization time of resources to reach the given time frame.

(iii) Consideration of different recycling paths within a cascaded utilization
In the case study, all mass flows enter the same recycling process. However, such a one-dimensional cascaded utilization is rather unlikely in reality. A diversion into different recycling processes is to be expected, especially for later cascade stages. For example, in Scenario I, after initial use as an SMC component, it could have been assumed that only part of the material is reused, while the other part is used for direct energy recovery. However, such an approach significantly increases the complexity of the model and the required data diversity.

(iv) Technological development
A study on cascaded utilization of a product system is carried out using state-of-the-art technology. The timeframe under consideration can be very long (e.g. 40 years). A further technological development is to be expected, which can significantly influence recycling processes. This uncertainty is relatively difficult to depict. Nevertheless, a combination with a risk model to illustrate the probability of further technological development would be conceivable.

(v) Amount of impact indicators
In the case study, only the cumulative energy and raw material demand were chosen as impact indicators. Even if some conclusions could be drawn this way, an examination of further impact indicators could enable an even more differentiated statement on the advantages and disadvantages of cascaded utilizations. Again, this increases the required data diversity significantly. The research effort always should be in line with the benefit of the studies.

5 Conclusion

Cascaded utilizations can be evaluated on the basis of indicators within a life cycle assessment. This paper discussed a method by which this can be achieved despite the multi-functionality of such systems. Based on a comparison of the cascaded utilization with an equivalence system, potential reductions can be identified, which then serve as a basis for comparison. The methodology was then tested in a case study of the cascaded utilization of a rotor blade. For four scenarios, potential reductions in the impact indicators of cumulated energy demand and cumulated raw material demand could be generated. In principle, a reduction in environmental impacts can always be expected for cascaded utilizations, provided that the environmental impact from the processing of the materials does not exceed the environmental impact from the production of the material to be substituted. Using the present method, however, reductions in environmental impacts could also be quantified and compared by considering impact indicators. This enables the identification of processes in which, for example, the resource or energy efficiency of the product system can be increased through cascaded utilization. Furthermore, the scenario for a cascaded utilization could be identified, where the greatest reductions of environmental impacts is to be expected. Such results can be used, for example, in environmental policy.

This paper also contributes to the further specification of the term "cascaded utilization" by establishing a cascade chain model that is coupled with the utilizations of the product system. This seems to be particularly relevant when looking at literature already published in this subject area. The term cascaded utilization is used in many different ways and sometimes remains unclear. A narrower definition seems desirable. The two-dimensional model proposed in Section 2.1 appears to be a useful visualization of cascaded utilizations and could be used for further work.

In the discussion of the method it became clear that comprehensive data diversity is required for the application of the methodology. In addition to data on the product system with cascaded utilization, data on the equivalence systems need to be provided as well. Furthermore, the result of such a study is highly dependent on the design of the scenarios and equivalence systems. This should always be presented as transparently as possible, so that a good traceability is guaranteed.

6 Zusammenfassung

Die Nutzung von Ressourcen in mehrfachen Kaskadenstufen kann die Ressourceneffizienz von Produktsystemen verbessern. Um verschiedene Kaskadennutzungen miteinander vergleichen zu können ist eine Bewertung des Beitrags der Kaskadennutzung durch quantifizierbare Indikatoren notwendig. In diesem Beitrag wird eine Methode diskutiert, mit der ein Vergleich der Kaskadennutzungen trotz der Multifunktionalität dieses Systems ermöglicht wird. Durch den Vergleich der Kaskadennutzung mit einem äquivalenten System können potenzielle Verbesserungen der Indikatoren identifiziert werden. Im Rahmen einer Fallstudie werden vier Szenarien für die mehrfache Verwendung eines Rotorenblatts durch eine Literaturanalyse entwickelt. Potenzielle Reduktionen in den Indikatoren Kumulativer Energiebedarf (CED) und Kumulativer Rohstoffbedarf (CRD) werden berechnet. Signifikante Reduktionen im Primärenergiebedarf können bei der Verwendung von recycleter Glasfaser in Sekundärprodukten sowie bei einer Nutzung des Rotorblattmaerials als Sekundärbrennstoff in der Zementindustrie beobachtet werden. Die höchste Reduktion im Rohstoffbedarf erfolgte durch das mechanische Recycling von Rotorblättern und Nutzung des Materials als Füllstoff. Im Hinblick auf die Anwendung der Methode wird deutlich, dass (i) die Auswahl des Äquivalenzsystems einen entscheidenden Einfluss auf die Ergbenisse hat und (ii) diese Herangehensweise eine hohe Datenverfügbarkeit und –diversität notwendig macht.

Acknowledgments. We thank our colleagues from the Research Group Cascade Use, in particular Dr. Alexandra Pehlken, who provided insight and expertise that greatly assisted the research.

7 References

[1] BMUB (2016) German Resource Efficiency Program II: Programm zur nachhaltigen Nutzung und zum Schutz der natürlichen Ressourcen. https://www.bmu.de/

[2] Klöpffer W, Grahl B (2009) Ökobilanz (LCA): Ein Leitfaden für Ausbildung und Beruf. Wiley-VCH, Weinheim, 1st ed., ISBN 978-3-527-32043-1

[3] Risse M, Richter K (2016) Nutzung nachwachsender Rohstoffe in Kaskaden – Ansätze zur lebenszyklusorientierten Bewertung der ökologischen und ökonomischen Effekte. uwf 24:63-68. doi:10.1007/s00550-016-0391-x

[4] Höglmeier K, Steubing B, Weber-Blaschke G, Richter K (2015) LCA-based optimization of wood utilization under special consideration of a cascading use of wood. J Environ Manage 152:158–170. doi:10.1016/j.jenvman.2015.01.018

[5] Paraskevas D, Kellens K, Dewulf W, Duflou J (2015) Environmental modelling of aluminium recycling: A Life Cycle Assessment tool for sustainable metal management. J Clean Prod 105:357–370. doi:10.1016/j.jclepro.2014.09.102

[6] Korhonen J (2000) Completing the industrial ecology cascade chain in the case of a paper industry - SME potential in industrial ecology. Eco-Management Audit 7:11–20. doi:10.1002/(SICI)1099-0925(200003)7:1<11::AID-EMA121>3.0.CO;2-C

[7] Höglmeier K, Weber-Blaschke G, Richter K (2014) Utilization of recovered wood in cascades versus utilization of primary wood—a comparison with life cycle assessment using system expansion. Int J Life Cycle Assess 19:1755–1766. doi:10.1007/s11367-014-0774-6
[8] Sirkin T, Houten M (1994) The Cascade Chain: A Theory and Tool for Achieving Resource Sustainability with Applications for Product Design. Resour Conserv Recycl 10:213–276. doi:10.1016/0921-3449(94)90016-7
[9] International Ogranization for Standardization (2006) ISO 14040:2006 Environmental management - Life cycle assessment - Principles and framework.
[10] International Organization for Standardization (2006) ISO 14044:2006 Environmental management -- Life cycle assessment -- Requirements and guidelines.
[11] BMWI (2916) EEG-Novelle 2016: Fortgeschriebenes Eckpunktepapier zum Vorschlag des BMWi für das neue EEG. https://www.bmwi.de/
[12] Seiler E. (2010) Recycling von Windkraftanlagen. Pflinztal, Internal report
[13] Albers H, Greiner S, Böhm A (2011) Recycling of Rotor Blades from Wind Turbines. In: Pehlken A, Solsbach A, Stenzel W (eds) Hanse-Studien 2011. pp. 67–75. ISBN 978-3-814-22283-7
[14] Tryfonidou R (2006) Energetische Analyse eines Offshore-Windparks unter Berücksichtigung der Netzintegration. Ruhr-Universität, Bochum
[15] Pickering SJ (2006) Recycling technologies for thermoset composite materials—current status. Compos Part A Appl Sci Manuf 37:1206–1215. doi:10.1016/j.compositesa.2005.05.030
[16] Beauson J, Lilholt H, Brøndsted P (2014) Recycling solid residues recovered from glass fibre-reinforced composites – A review applied to wind turbine blade materials. J Reinf Plast Compos 33:1542–1556. doi:10.1177/0731684414537131
[17] DeRosa R, Telfeyan E, Mayes S (2004) Expanding the use of recycled SMC in BMCs. Presentation, Global Plastics Environmental Conference 2004 (GPEC)
[18] Martens H, Goldmann D (2016) Recyclingtechnik: Fachbuch für Lehre und Praxis. Springer, Wiesbaden, 2nd ed., ISBN 978-3-658-02786-5, doi:10.1007/978-3-658-02786-5
[19] Onwudili JA, Miskolczi N, Nagy T, Lipóczi G (2016) Recovery of glass fibre and carbon fibres from reinforced thermosets by batch pyrolysis and investigation of fibre re-using as reinforcement in LDPE matrix. Compos Part B Eng 91:154–161. doi:10.1016/j.compositesb.2016.01.055
[20] Pickering SJ, Kelly RM, Kennerley JR, Rudd CD, Fenwick NJ (2000) A fluidised-bed process for the recovery of glass fibres from scrap thermoset composites. Compos Sci Technol 60:509–523. doi:10.1016/S0266-3538(99)00154-2
[21] Woidasky J (2013) Weiterentwicklung des Recyclings von faserverstärkten Verbunden. In: Thomé-Kozmiensky KJ, Goldmann D (eds) TK Verlag 2013. pp. 241-259. ISBN 978-3-935317-97-9
[22] Friedel A (1999) Einfluss der Produktgestalt auf den Energieaufwand beim Recycling mechanischer Bauteile und Baugruppen. Springer, Berlin, ISBN: 978-3-642-47976-2
[23] Umweltbundesamt (1999) Basisdaten und Methoden zum kumulierten Energieaufwand. https://www.umweltbundesamt.de/
[24] European Commission (2006) Reference Document on the Best Available Techniques for Waste Incineration. Report.
[25] VDI (2012) Richtlinie 4600: Kumulierter Energieaufwand (KEA) - Begriffe, Berechnungsmethoden.
[26] VDI (2016) Richtlinie 4800: Blatt 2 Ressourceneffizienz - Bewertung des Rohstoffaufwands.

Development and Application of Metrics for Evaluation of Cumulative Energy Efficiency for IT Devices in Data Centers

Fernando Peñaherrera[1] and Katharina Szczepaniak[2]
[1]Carl von Ossietzky University, Oldenburg, 26111, Germany
[2]Technische Universität Hamburg, Hamburg, 21073, Germany
fernando.andres.penaherrera.vaca@uni-oldenburg.de

Abstract

This paper develops and evaluates a set of metrics to evaluate the holistic energy efficiency of data centers. The increasing energy consumption from data centers led to the development of various metrics to monitor energy efficiency and environmental impact. However, these metrics overlook stages outside the operation phase, and often overlook energy transformation losses outside of the boundaries of the data center.

The Cumulative Energy Efficiency (CEE) and Cumulated Performance Efficiency (CPE) are metrics developed considering criteria for indicators for the evaluation of resource efficiency, while also taking into account aspects of sustainability and primary resource depletion. The metrics are calculated using Cumulative Energy Demand as resource indicator, which is analyzed through a Life Cycle Assessment of products. The useful energy and performance is then compared to this resource depletion indicator. The metrics are then tested with a case study of a server used in a data center. The results indicate a CEE=0,260 and a CPE=789 ops/J. When comparing these values to other well established metrics, the developed metrics account for embodied energy and energy transformation losses for the whole energy supply chain. A sensitivity analysis performed showed an independence of these metrics when considering changes in server load and idle times, and parallel behavior when considering variations in energy conversion efficiency inside the data center boundaries. These metrics allow a complete monitoring of the energy performance of IT devices for data centers from cradle to grave. The metrics are to be tested in additional case studies and can be aggregated for analysis of whole infrastructures. These can be used by decision makers for improvements in design, operation, and End-of-Life strategies. The dependence of the proposed metrics on other natural resources, such as raw materials, is also to be analyzed.

1 Introduction

A data center is a building or space where computing capacities (server, storage, network) and the infrastructure for its operation (power supply, climatization, security) are centrally located [1–4]. A data center handles IT load in the form of applications or services for processing, storage, and sharing of data [2]. The size of the IT load capacity depends on the type and number of installed servers. This determines the energetic dimensioning of the supporting devices [5].

Data centers consume high levels of energy to operate the IT equipment and to extract the heat they produce. Direct electricity used in data centers reached 2% of total world electricity consumption in 2014, due to an increasing number of implemented data centers [4,6–8]. In 2011, the electricity consumption of data centers in the European Union (EU) was 52 TWh, and 70 TWh will be necessary for their operation by 2020 [9,10]. The workload of data centers by 2020 will be triple that of 2012 [9,11]. The Borderstep Institute calculated that data centers are responsible for 1,8% of total electricity consumption in Germany [12–14]. Data centers have thus an important role in the implementation of the EU climate protection goals and in the achievement of the efficiency goals of the German Federal Government [9]. For Germany, the energy consumption of data centers is forecasted to reach over 14 TWh in 2020 and 16,4 TWh in 2025 [3].

Because of their reliance on power, and associated operational cost, metrics developed for data centers use operational efficiency as a proxy for sustainability [15]. Savings in energy consumption in data centers are achieved through virtualization, improved hardware, and optimization in climatization [14]. Despite this, the electricity consumption of data centers in Germany increased 3% to 12 TWh in 2015, with an accelerated increase in their energy demand [3,8,16]. Additional to the issues of energy efficiency the questions of material efficiency of data centers should be regarded. The development of appropriate end of life (EOL) strategies for data center components should be enforced [17]. A complete evaluation of energy consumption during the whole lifecycle of these components is required to analyze such strategies.

To evaluate the complete energy consumption of data center equipment, this paper develops a metric for resource efficiency which allows a comparison of products in a data center during their whole lifecycle. Due to the vast variety of metrics for data centers, a survey is performed to assess the requirements of such metrics and to find where gaps in sustainability criteria can be addressed. Then, considering the requirements for evaluation of holistic assessment of energy efficiency, a set of metrics is formulated. These metrics are then tested with a case study performed on a data center IT device. The energy demands are evaluated using a Life Cycle Analysis (LCA), where different scenarios on performance, lifetime and infrastructure efficiency are modelled to observe impacts on the evaluated metrics [18].

A. Pehlken et al. (Eds.), *Cascade Use in Technologies 2018*,
https://doi.org/10.1007/978-3-662-57886-5_17

2 Metrics for Data Center Resource Efficiency

Metrics as management tools serve to monitor the effectiveness of an organization to ensure clear operational and environmental conditions [19,20]. Metrics are useful to measure IT performance and productivity and to identify problem areas. There are several metrics to measure and monitor different aspects of data centers. Green metrics measure the environmental impact of a data center and its components [1]. Common measurements used in data center management are energy consumption, energy cost, or carbon emissions. The increasing energy demand of data centers motivated the development of metrics for the depiction of energy efficiency of the infrastructure [21,22]. Other approaches include monitoring IT performance [23,24]. However, these metrics overlook the impacts of stages other than the operation phase.

2.1 Requirements of a Metric

A metric measures a characteristic of an object or system and includes a procedure or methodology for making this measurement [25]. Effective metrics should be intuitive, accurate, granular, and representative of the system [26]. A metric for resource efficiency should quantify useful work relative to the amount of resources consumed [25]. Criteria for metrics for the evaluation of resource efficiency are:

Independence: In a complex system like a data center, dependencies between the functional systems are present. These interactions should be presented and quantified in the metrics [27].

Completeness: Metrics should be able to analyze individual functional systems. They should map all the environmental essential aspects for a data center.

Uniformity: The data collection methods in a metrics system should be similar for metrics of equal importance, avoiding combination of exact measurements with data from estimates [28].

Quantitative: Metrics should be measurable so that they can be acted upon.

Comparability: Reported data shall be in a comparable format, expressed in absolute terms [29].

Clearness: The content shall be clear and have a traceable optimization direction.

Optimizable: The metrics shall provide information on improvement approaches [18].

Metrics should address the most significant environmental aspects that can be influenced [30,31]. Metrics shall follow the general guidelines of the ISO 14031 standard and link them to the corresponding environmental aspects [30]. Natural resources under consideration are energy resources. An indicator associated with these resources is the Cumulative Energy Demand (CED) [32].

Existing Metrics for Data Centers

The resource efficiency of a data center is defined as the ratio of IT performance to the use of natural resources [18]. To achieve more energy efficiency and sustainability in a data center, metrics to monitor efficiency need to be applied [7]. Various metrics exist on different dimensions: energy efficiency, cooling, greenness, performance, thermal and air management [1]. Most metrics focus on the efficient use of individual resources during the operation of a data center, helping to reduce operational expenditure [33]. Standards for data center metrics already under development and can be found in the ISO 30134, ISO 21836, and EN 50600 [18].

Metrics for Resource Efficiency in Data Centers

Different metrics have been developed to determine a data center's energy efficiency [9]. Most of the metrics developed relate to individual systems in the data center: cooling, UPS, servers, or type of energy use. Other metrics measure emissions, water consumption, or the disposal efficiency [1]. Metrics that monitor the whole data center infrastructure are the Power Usage Effectiveness (PUE), or the Corporate Average Data Center Efficiency. Other metrics consider not only the energy but also the work and productivity of the data center, such as the Data Center Energy Productivity [34] or the Data Center Performance per Energy [35]. The Load Dependent Energy Efficiency monitors the relationship between Performance and consumed power [2]. Table 1 provides an overview of data center metrics for energy efficiency.

The range of metrics available shows that no single metric entirely fits all purposes [6].There is a need for performance metrics that better capture the holistic efficiency of a given data center, considering aspects of sustainability. Combinations of metrics are used to offset these limitations [28]. Some metrics evaluate the overall operational efficiency of a data center. Others evaluate the IT performance of the equipment [1,36]. The three dimensions of computing, data storage and data transfer can also be depicted with separate indicators [18,37].

Table 1. Overview of metrics for data center resource efficiency

Metric	Name	Source
EDE	Electronics Disposal Efficiency	[1]
ERE	Energy Reuse Effectiveness	[38]
ERF	Energy Reuse Factor	[38]
GEC	Green Energy Coefficient	[39]
GUF	Grid Utilization Factor	[40]
MRR	Material Recycling Ratio	[41]
WUE	Water Usage Energy	[42]
TCE	Technology Carbon Efficiency	[43]
FpW	FLOPS per Watt	[44]
PUE	Power Usage Effectiveness	[45]
SI-EER	Site Infrastructure Energy Efficiency	[16]
DCiE	Data Center Infrastructure Efficiency	[45]
DCeP	Data Center Energy Productivity	[46]
ScE	Server Compute Efficiency	[47]
DCcE	Data Center Compute Efficiency	[47]
DC-EEP	DC Energy Efficiency and Productivity	[16]
CUE	Carbon Usage Effectiveness	[48]
LDEE	Load Dependent Energy Efficiency	[2]
CCF	Cooling Capacity Factor	[9]
COP	Coefficient of Performance	[49]
$pPUE_{cool}$	Partial PUE Cooling	[50]
$pPUE_{UPS}$	Partial PUE UPS	[50]
pPUE	Partial PUE	[50]
CADE	Corporate Average Datacenter Efficiency	[51]
SPECpower	Standard Performance Evaluation	[52]
U_{server}	Server Utilization	[37]
$U_{network}$	Network Utilization	[37]
$U_{storage}$	Storage Utilization	[37]

Currently, two well-established metrics are used to determine energy efficiency for data centers and systems: Power Usage Effectiveness (PUE) and FLOPS per Watt (FpW). PUE focuses on operational efficiency, using it as a proxy for sustainability. It is calculated as the ratio of the energy usage of a data center IT equipment divided by the total facility energy usage [6]. Partial PUE (pPUE) metrics focuses particular subsystems [53]. The Green Grid consortium proposed the Data Center Infrastructure Efficiency (DCiE), which is the inverse of the PUE [1,54]. FLOPS per Watt characterizes the energy efficiency of a system by comparing the useful computing work to the energy consumption of the data center.
These metrics miss impacts that are embodied in the facility due to energy consumed and emissions created during the manufacturing and disposal of the data center and their components. Whether the servers themselves are operated efficiently cannot be determined via the PUE. Comparisons between data centers using PUE are most often not representative of the actual situation [1]. By only considering one environmental issue in one stage of the data center lifetime, the effects that improving the efficiency of one issue has on another is omitted [33,55].
To address these issues, this paper introduces the metrics Cumulated Energy Efficiency (CEE) and Cumulated Performance Efficiency (CPE). CEE calculates the useful energy consumption during the operation phase and relates it to the cumulated energy demand (CED) during its lifecycle. Similarly, CPE relates the total performance to the CED.

Cumulated Energy Demand

When discussing sustainability aspects, indicators suitable for measuring the use of natural resources need to be applied [32]. UBA (2002) identified energy as a natural resource and developed a set of indicators considering the definitions of natural resources for Germany. Energy as a resource indicator is represented in the German sustainability strategy in the form of cumulative primary energy demand (CED) [32,56].
Energy resources are natural resources used to generate energy services. They include material fuels (fossil fuels, biomass, and uranium), energy mediated by substances (wind, hydropower, geothermal), and others such as radiation [32]. CED also includes the energy expenditure contained in phases outside the use phase of products. This energy indicator can be calculated through LCA of products.

2.2 Cumulative Energy Efficiency and Cumulated Performance Efficiency

The presented metrics consider operational energy efficiency of the data center or a partial system of it, and overlook energy efficiency of individual IT components [9]. Borderstep (2017) [9] addresses the difficulties in identifying a metric for objective comparison of the energy efficiency of different data centers. The observed metrics overview the energy required in phases outside of the operational phase. Moreover, the CED is also omitted, focusing mostly in the final energy, overlooking energy transformation losses. Although some metrics consider the EOL of devices in a data center, the impact that reuse or recycling strategies can have in the CED of the device are unknown.

A sustainability performance metric that considers energy efficiency needs to compare the useful work done by a system to the total resource consumption [2,57]. To evaluate primary energy consumption in data center, an aggregated inventory of the different materials and processes involved across the lifecycle of the system needs to be compiled. This inventory is then translated into environmental impacts through a set of impact factors to calculate the resource depletion [58,59].
To estimate the output of an IT device in a data center, power–performance benchmarks are employed to analyze productivity in the operational phase of the data center. Therefore, a metric that quantifies the useful work that a piece of equipment produces and then relates it to the amount of resources it consumes provides the correct tool [25]. Using the Load-Power relation in these devices, the amount of useful work can be related to the energy consumption [24,52]. Since the productivity of the data center equipment can be related to its power ($P_i(L)$), the useful work (E_{useful}) of a device during its lifetime ($a_{i,LT}$) can be estimated with:

$$E_{useful} = \sum_{i=1}^{n} P_i(L) \cdot a_{i,LT} \tag{1}$$

The Cumulated Energy Efficiency (CEE) is proposed as a metric to evaluate the energy efficiency of a data center device during its lifetime, by relating the useful work during its operational phase to the CED during its lifetime:

$$CEE = \frac{E_{useful}}{CED} \tag{2}$$

If the number of operations per second and its relation to the power consumption is known, the Cumulated Performance Efficiency (CPE) can be expressed as operations per primary energy input:

$$CPE = \frac{Operations}{CED} \tag{3}$$

These metrics normalize the data for comparison between IT devices in a data center and consider the utilization of these devices. These metrics are focused on the different subsystems and their components, such as servers, network, and storage units. The additional infrastructure is dedicated to support the operation of these components. When aggregated for different subsystems, they allow a comparison between whole infrastructures.
In the following section the applicability of these metrics is evaluated through a case study of a specific device in a data center.

3 Application of CEE and CPE

3.1 Data Center Infrastructure and Components

A data center encompasses all the facilities and infrastructures for power distribution and environmental control together with the necessary security required to provide the desired service availability A data center essentially consists of three sub-systems, in addition the building facility: IT infrastructure, power infrastructure, and cooling infrastructure [58,60]. The IT provides the data processing capabilities and contributes to 50-80% of a data center energy demand. The IT infrastructure consists of three main components: servers, networking equipment and storage devices [58].
Due to their high energy consumption and short lifetime, the proposed indicators are applied first to a server in a data center. Servers come as tower, rack-optimized, multi-node, or mainframe servers. Servers consist of mainboards, central processing units (CPU), random access memory (RAM), hard disk or solid-state drives (HDD/ SSD), network interfaces (NIC), fans and power supply units (PSU). CPU, RAM and PSU are the main power consumers of servers [61,62].

3.2 Life Cycle Assessment for Cumulative Energy Demand Calculation

The environmental footprint of ICT systems consists mostly of impacts related to the measurable consumption of materials and energy in its lifecycle [58]. To evaluate the CED a lifecycle approach is required where the performance of a device from its manufacturing to eventual disposal is considered [15].
LCA is a structured, comprehensive, and internationally standardized method. It quantifies all relevant emissions and resources consumed and the related environmental impacts and resource depletion associated with any product. The ISO 14040 and 14044 standards provide the framework for LCA [59,63,64]. Through the compilation of an inventory, an LCA looks at the products and processes contained within a system, from the extraction of raw materials through manufacturing, transportation, operation, and eventual disposal [15]. The current state of LCA methodologies, the lack of applicable primary and secondary data for assessing data center components and systems, and the complexity of a data center create serious difficulties in performing LCA. Equipment or systems can be removed from the LCA scope if it is proven that the impact is negligible (less than 2%) compared with the whole data center [60].

Goal and Scope Definition

The purpose is to evaluate the CED of a server (PowerEdge M710) during its lifetime. The functional unit is a unit of this product during its lifetime. To calculate the required indicators, data on product lifetime, parts, composition, use phase energy, and recycling factors is required. Data for the inventory analysis is measured, calculated, derived from other sources, or as proxy data from similar activities [60]. A model is built in openLCA 1.7 using the ecoinvent 3.4 database.

Inventory Analysis

Lifetime

Expected lifetime for high-end servers ranges from 5 to 8 years [60,65]. Typically, servers are decommissioned after 5 years due to expiration of warranty. The operation of a server in a data center is characterized by a continuous operation mode, with unavailability typically lower than 12h/a [66,67].

Parts and Components

A disassembly of the unit was performed to obtain a component list with the relative weight of the parts. Szczepaniak [69] conducted a detailed analysis of the material and elemental composition of the unit. Within the materials analyzed, base metals (Fe, Cu, Al, Zn) and plastic (for molded parts) are assigned as base materials. Precious metals considered are Au and Ag. Critical Raw Materials (CRM) are considered as per the definition of the European Union [68]: platinum group metals (Pd, Pt), light rare earth elements (Nd), heavy rare earth elements (Y), and additionally Ta, Wo, Ga, In, Sn, Be, Co, Ge are expected to be used in data center devices. Pb and Li are categorized as strategic valuable metals due to their use in batteries [69]. Estimations on the required transportation of the parts are performed.

Useful Energy

To model the energy consumption of a ICT device in a data center, it is important to know if a component is load-adaptive, or if the energy demand is approaching static [5].

The Standard Performance Evaluation Corporation (SPEC) has developed standardized performance benchmarks for computer systems [23]. The benchmark SPECpower_ssj2008 determines the power requirements of a server at different load levels while measuring the power consumed [25]. The computing power is estimated by determining the performance of the main processors of the server and multiplying it by their utilization in operation [18,70]. Due to the limited data on load rates for servers in data center, this study generalizes server utilization a assuming 80% utilization 50% of the time and idled the remaining time [71]. This allows estimating the useful energy consumed and useful work performed.

Power Supply Infrastructure and Energy Losses

The power infrastructure is responsible for providing uninterrupted, conditioned power at the correct voltage and frequency to the IT equipment. From the utility feed, the power goes through step down transformers, transfer switches, uninterruptible power supplies (UPS), PDUs, and finally to rack power strips. The UPS conditions power and provide backup power in case of short outages. Generators may be used for longer outages [58].

Power provisioning usually consumes 5-20 % of a data center total power [62]. IBM indicates that the overall energy supply and conversion efficiency in a data center are typically at 73% [72]. Losses primarily occur through transformers, UPS, and PDU, whereas losses of cabling are small [62,73]. A high-quality high-performance UPS achieves efficiencies up to 96%, and smaller systems reach at least 90% efficiency [74].

Climatization

IT room air conditioning is responsible for removal of the heat generated by the IT equipment. It is assumed that all electric power consumed by the servers is transformed into heat and must be evacuated [5]. The estimation of the energy required for climatization considers the use of cooling units which use refrigeration cycle. The power consumption ($P_{chiller}$) is dependant on the amount of heat removed ($\dot{Q}_{removed}$) and on the coefficient of performance (COP):

$$P_{chiller} = \frac{\dot{Q}_{removed}}{COP} \tag{4}$$

Typical values of COP at operational conditions are around 2,0 [2,5]. This allows estimating the required energy consumption for heat extraction.

End of Life

End of Life for data center components is not documented. Most of the devices are disposed as Waste Electrical and Electronic Equipment (WEEE), and there are currently no strategies for reuse after their service phase. Recycling of WEEE focuses on the recovery of valuable metals, and the recovery of energy through incineration. The simplest form of recycling is closed-loop recycling: the secondary good is shunted back to an earlier process in the same system where it directly replaces input from primary production of the material [63]. It is expected all metals to be recycled and all non-metals assumed to be incinerated or recycled, in a 50%-50% share [75]. UBA (2018) assumes a required 200 km transportation, manual disassembly (without energy consumption), shredding of WEEE, and incineration of other components. Scrap from housing is modeled as iron or aluminum scrap.

LCA Model

The ecoinvent 3.4 (2017) database is used to construct the model. Due to the absence of information regarding manufacturing processes, these different parts were modelled having as a source the general manufacturing techniques available on the database. The different losses were modelled as inefficiencies of the electricity supply equipment. A detail on the different flows and matching processes is described in Table 2. Figure 1 represents the model built in openLCA 1.7.

Figure 1. LCA Model

Table 2. Inventory list for LCA

Process	Value	Unit	Modelled as
Parts Manufacturing [69]			
Hard Drives	0,284	kg	hard disk drive, for laptop computer
Processor	0,032	kg	printed wiring board, mounted mainboard
Casing	6,237	kg	steel, unalloyed
SD cards	0,003	kg	integrated circuit, logic type
Mainboard (Q-logic)	0,056	kg	integrated circuit, logic type
SD card reader	0,023	kg	integrated circuit, logic type
Mainboard	0,064	kg	printed wiring board (PWB), mounted mainboard,
Mainboard (network)	0,051	kg	printed wiring board, for power supply unit
Heat Sink (Al)	0,033	kg	section bar extrusion, aluminum
Backplane (SATA)	0,047	kg	printed wiring board, for surface mounting
Motherboard	1,458	kg	printed wiring board, mounted mainboard
PCB (HDD / battery)	0,048	kg	printed wiring board, for surface mounting
Battery (Li)	0,003	kg	battery, Li-ion, rechargeable
Battery (Li)	0,050	kg	battery, Li-ion, rechargeable
Cables	0,003	kg	cable, unspecified
Plastic	0,207	kg	polystyrene, general purpose
Main Memory	0,019	kg	integrated circuit, memory type
Assembly			
Transport	1,7	ton*km	transport, freight, lorry- GLO
Assembly Energy	8 787	kWh	electricity, medium voltage - CN
Transport	195	ton*km	transport, freight, sea, transoceanic ship | - GLO
Transport	1,7	ton*km	transport, freight, lorry - GLO
Operation			
Energy	174,5	MWh	electricity, low voltage - DE
End of Life			
Transport	1,7	kg*km	transport, freight, lorry - GLO
WEEE	4,31	kg	treatment of WEEE, shredding - GLO

3.3 Results on CED and Evaluation of Metrics

The results on CED indicate a strong prevalence of the use phase in energy consumption, accounting for more than 90% of the total CED (Figure 2). From the other phases, the parts production and assembly phase accounts for almost all the remaining CED. Ramesh et al. and Whitehead et al. also indicate that operational energy accounted for 80-90% of the impacts [15,76] .

Figure 2. Results on CED for the server model PowerEdge M710.

For calculation of the CEE and CPE, the amount of useful energy consumed by the server (included in idle states) is divided by the CED. Additionally, when considering the total operations executed, the CPE can also be calculated as operations per unit of energy:

$$CEE = 0{,}260; CPE = 789\frac{ops}{J} \tag{5}$$

To validate these results, a comparison with similar efficiency indicators using the same values indicates a PUE of 2,0, indicating a DCiE of 0,5. The average Performance to Power ratio is 3 162 $\frac{ops}{J}$. The CEE considers inefficiencies on energy conversion losses and for the embodied energy of the device.

A sensitivity analysis indicates the influence of the various factors considered for this evaluation. Parameters such as operation time, load, idle time, and efficiencies of the power supply infrastructure are modified to study the impact on the metrics (Figure 3).

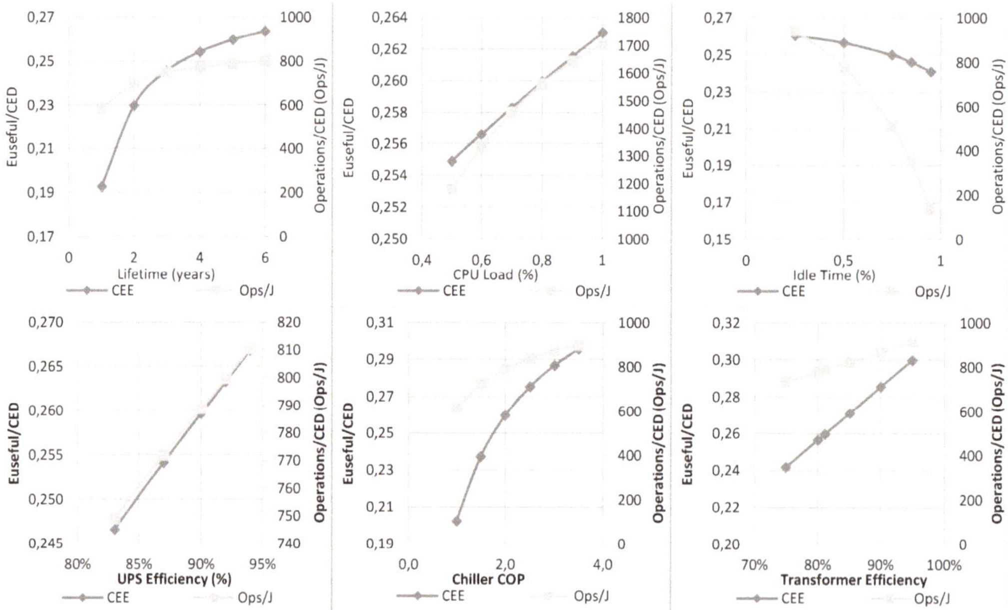

Figure 3. Results on CEE and CPE for different configurations

4 Results Discussion and Interpretation

The application of CEE and CPD as indicators provide information of different aspects of the IT device analyzed. They provide a general overview of the overall efficiency of the utilization of energy for this application. When coupled with the performance, CPD provides information on the resource efficiency per operation in the unit.
The results on CED agree with the observations of embodied energy and operational energy, when considering a lifetime of 5 years [15]. When changing this value, the embodied energy becomes more representative as the lifetime is reduced.
No single metric will be able to cover all dimensions of data center performance [1,6,77]. The CEE allows a direct comparison between servers and IT devices used in data centers, while considering all infrastructure and the different stages in its lifecycle, and the different conditions of operation. This indicator can also be applied to analyze impacts of virtualization and other location-based strategies.
Since the use of IT power consumption as a proxy for IT productivity presents issues of linearity with performance [78], the CEE and CPE allow a distinction between load and performance. The results indicate different behavior of these metrics when considering changes in idle time and in load. In other cases, the indicators move in parallel.
Given that some of the data was secondary, the quality of the results is affected. Most of the data used for manufacturing processes was matched to generalized processes with geographical and temporal similarities, which can provide uncertainties of 20% to 30% in estimates on a single, high-volume server system [60]. The different parts and components where modelled after similar processes for laptops and hard disk components, which lead to higher uncertainties for the manufacturing stage. However, the most influential stage, the use phase, has in general good data quality, since the energy consumption in data centers is well documented, and the energy source was modelled after the German energy mix. Data on EOL is highly uncertain, since a 50% recycling share was assumed, but the recycling quota of WEEE can be below 15% [79,80], and losses in the recycling processes were not considered. The sensitivity analysis indicates where more precision in input data is required. Precise data on operational lifetime and in load during use phase is required. COP of the chiller and the potential use of free cooling have a major influence in the value of the metrics analyzed.

5 Conclusions and Future Work

The CEE and CPE as indicators for holistic energy efficiency of IT devices in data centers were introduced. These indicators use a complete LCA of these devices to evaluate the CED. By matching this resource indicator to the useful energy and useful work performed, this metric allows comparison of different operation conditions, energy supply infrastructure, and different devices used for the same application.
These metrics have relation to other well stablished metrics such as the PUE or DCiE. However, by assessing the overall energy consumption during the different stages of the lifecycle, they allow a holistic observation of the energy performance during the different life stages and under different working conditions. These metrics have potential to be tested into different case studies. The use of these metrics as indicators of energy efficiency in a bigger scale needs to be evaluated.
The use of these metrics can allow taking decisions which lead to overall savings of energy and optimization of energy performance. With the inclusion of EOL in the analysis of the metric, proper strategies for reuse, replacement, or recycling of devices can developed, which also consider optimization of resources.
Measurement of resource efficiency in a data center has the possibility of becoming the standard [18]. The dependence of the energy efficiency and other environmental indicators, such as Cumulated Material Demand (CMD), and of critical material depletion, needs to be studied. While the correlation of CED with other indicators has been analyzed [18,32], the impact of different operational strategies on the indicators is still unclear [17]. The application of the proposed metrics in combination with other indicators can indicate an interdependence between performance, energy and material demands. The European economic systems are dependent on natural resources (raw materials and energy), and conflicts or synergies between these natural resources are not well stablished [32,81].
Criteria on independence of these metrics are to be tested by comparing it to other environmental resource indicators, such as material depletion. The metrics have also the potential to be aggregated to analyze complete systems, and their normalized format allows comparison between systems and infrastructures. The data collection methods need improvement, since a mixture of primary, secondary and proxy data was used. Data quality and the impact on uncertainty were disregarded in this analysis.
The proper assessment of these metrics needs to be accompanied by an evaluation of the uncertainty, so that improvements on data quality are addressed. Improvements are required on data of manufacturing processes for part and components used in IT devices for data centers and on recycling quotas for these devices. EOL was modelled as WEEE, but these components are to be collected separately. The sensitivity analysis allows indicating where actions can be taken for improvement of efficiency, both in energy and in performance efficiency.

Zusammenfassung

In diesem Beitrag wird eine Reihe von Metriken für die umfassende Bewertung der Energieeffizienz von Rechenzentren entwickelt und evaluiert. Der steigende Energieverbrauch von Rechenzentren hat zur Entwicklung verschiedener Metriken zur Kontrolle des Energieverbrauchs und der Umweltauswirkungen von Rechenzentren geführt. Diese übersehen jedoch häufig die Stufen außerhalb der Nutzungsphase und beachten Energieumwandlungsverluste außerhalb der Grenzen des Rechenzentrums nicht. Die entwickelten Metriken Kumulative Energieeffizienz (CEE) und Kumulierte Leistungseffizienz (CPE) berücksichtigen nicht nur Indikatoren für die Bewertung der Ressourceneffizienz, sondern auch Aspekte der Nachhaltigkeit und Erschöpfung der primären Ressourcen. Die Metriken werden unter Verwendung des kumulativen Energiebedarfs als Indikator für Ressourcen berechnet, der wiederum durch eine Lebenszyklusanalyse von Produkten analysiert wird. Die nutzbare Energie und Leistung werden dann mit dem Indikator der Ressourcenerschöpfung verglichen. Die Metriken werden anhand einer Fallstudie eines Servers in einem Rechenzentrum getestet, wobei die Ergebnisse eine CEE = 0,260 und eine CPE = 789 ops/J ergeben. Ein Vergleich dieser Werte mit anderen etablierten Metriken zeigt, dass die entwickelten Metriken die enthaltene Energie und Umwandlunsgverluste entlang der gesamten Wertschöpfungskette ausweisen. Eine durchgeführte Sensitivitätsanalyse demonstrierte die Unabhängigkeit der Metriken in Bezug auf Server Load und Leerlaufzeit, sowie ein paralleles Verhalten bei Variationen in Umwandlungseffizienz innerhalb der Grenzen des Rechenzentrums. Die entwickelten Metriken ermöglichen daher eine vollständige Kontrolle der Energieleistung von Geräten in Rechenzentren von der Wiege bis zur Bahre. Die Metriken sollen in zusätzlichen Fallstudien getestet werden und können für die Analyse ganzer Infrastrukturen aggregiert werden. Diese können von Entscheidungsträgern für Verbesserungen in Bezug auf Design, Betrieb und End-of-Life Strategien genutzt werden. Die Abhängigkeit der Metriken von anderen natürlichen Ressourcen, wie Rohstoffen, sollte ebenfalls analysiert werden.

Acknowledgments. This research was done as part of the TEMPRO Project, financed by the German Federal Ministry for Economic Affairs and Energy (BMWi), funding number 03ET1418A.

6 References

[1] Reddy V, Setz B, Subrahmanya G, Gangadharan, G, Aiello, M. Metrics for Sustainable Data Centers. IEEE Transactions on Sustainable Computing. 2017; Vol. 2, N° 3:290–303.

[2] Schlitt D. Entwicklung einer auslastungsabhängigen Energieffizienzmetrik für Rechenzentren: Dissertation zur Erlangung des Grades eines Doktors der Naturwissenschaften. Oldenburg: Oldenburg University; 2016.

[3] Hintemann R. Energy consumption of data centers continues to increase: 2015 update. Berlin, Germany; 2015. Available from: https://www.borderstep.de/wp-content/uploads/2015/01/Borderstep_Energy_Consumption_2015_Data_Centers_16_12_2015.pdf (accessed: 29/06/2018].

[4] Koomey J. Worldwide Electricity Used in Data Centers. Environmental Research Letters. 2008; Vol. 3, N° 3:1–8. Available from: http://iopscience.iop.org/article/10.1088/1748-9326/3/3/034008/pdf [accessed: 29/06/2018].

[5] Janacek S. Identifikation von Freiheitsgraden und Wechselwirkungen in Rechenzentren unter Betrachtung elektrischer und thermischer Energie: Dissertation zur Erlangung des Grades eines Doktors der Naturwissenschaften. Oldenburg: Oldenburg University; 2017.

[6] Dimension Data. The Relationship Between Data Centre Strategy and Energy Efficiency: White Paper. London, UK; 2014. Available from: https://www.dimensiondata.com/zh-CN/Pages/Profile%20Boxes/The-relationship-between-data-cantre-strategy-and-energy-efficient-data-centres.aspx [accessed: 13/06/2018].

[7] Jamalzadeh M, Behravan N. An Exhaustive Framework for Better Data Centers' Energy Efficiency and Greenness by Using Metrics. Indian Journal of Computer Science and Engineering. 2012; Vol. 2, N° 6:813–822.

[8] Hintemann R. Trotz verbesserter Energieeffizienz steigt der Energiebedarf der deutschen Rechenzentren im Jahr 2016. Berlin, Germany; 2017. Available from: https://www.borderstep.de/wp-content/uploads/2017/03/Borderstep_Rechenzentren_2016.pdf [accessed: 13/06/2018].

[9] Borderstep. Aktuell genutzte Kennzahlen und Indikatoren zur Energieeffizienz in Rechenzentren: Arbeitspapier zu AP 3 „Informations- und Bewertungsmodelle für die Energieeffizienz in Rechenzentren" im Projekt TEMPRO. Berlin, Germany; 2017.

[10] Prakash S, Baron Y, Ran L, Proske M, Schlösser A. Study on the Practical Application of the New Framework Methodology for Measuring the Environmental Impact of ICT: Cost/benefit analysis – SMART 2012/0064. Brussels: European Commission; 2014.

[11] Cisco. Cisco Global Cloud Index: Forecast and Methodology 2012-2017. San Jose, CA, USA; 2013. Available from: https://www.slideshare.net/CiscoSP360/cisco-global-cloud-index-forecast-and-methodology-20122017 [accessed: 26/06/2018].

[12] Hintemann R, Fichter K. Server und Rechenzentren in Deutschland im Jahr 2012. Berlin, Germany; 2013 Available from: http://www.borderstep.de/pdf/Kurzbericht_Rechenzentren_in_Deutschland_2012__09_04_2013.pdf [accessed: 13/06/2018].

[13] Hintemann R. Consolidation, Colocation, Virtualization, and Cloud Computing: The Impact of the Changing Structure of Data Centers on Total Electricity Demand. In: Hilty, L, Aebischer, B: ICT Innovations for Sustainability; 2015; p. 125–136.

[14] Borderstep. Rechenzentren in Deutschland: Eine Studie zur Darstellung der wirtschaftlichen Bedeutung und der Wettbewerbssituation. Berlin, Germany; 2014. Available from: https://www.bitkom.org/noindex/Publikationen/2014/Studien/Studie-zu-Rechenzentren-in-Deutschland-Wirtschaftliche-

Bedeutung-und-Wettbewerbssituation/Borderstep-Institut-Studie-Rechenzentren-in-Deutschland-05-05-20141.pdf [accessed: 29/06/2018].
[15] Whitehead B, Andrews D, Shah A. The Life Cycle Assessment of a UK Data Centre. The International Journal of Life Cycle Assessment. 2015; Vol. 20, N° 3:332–349.
[16] Brill K. Data Center Energy Efficiency and Productivity: White paper. Santa Fe, NM, USA; 2007. Available from: http://large.stanford.edu/courses/2017/ph240/yu2/docs/brill.pdf [accessed: 13/06/2018].
[17] Umweltbundesamt. Materialbestand der Rechenzentren in Deutschland - Eine Bestandsaufnahme zur Ermittlung von Ressourcen- und Energieeinsatz. Berlin, Germany: Umweltbundesamt; 2010.
[18] Umweltbundesamt. Kennzahlen und Indikatoren für die Beurteilung der Ressourceneffizienz von Rechenzentren und Prüfung der praktischen Anwendbarkeit. Berlin, Germany; 2018.
[19] Campos L, Melo H, Verdinelli M, Cauchick M. Environmental Performance Indicators: A study on ISO 14001 certified companies. Journal of Cleaner Production. 2015; Vol. 99:286–296.
[20] Boog E, Bizzo W. Utilização de Indicadores Ambientais como Instrumento para Gestão de Desempenho Ambiental em Empresas Certificadas com a ISO 14001. In: SIMPEP: X Simpósio de Engenharia da Produção. Bauru, SP, Brazil; 2003.
[21] The Green Grid. Green Grid Metrics: Describing Data Center Efficiency: Technical Committee White Paper. Beaverton, OR, USA; 2007. Available from: https://leonardo-energy.pl/wp-content/uploads/2018/03/Green_Grid_Metrics.pdf [accessed: 29/06/2018].
[22] Anderson D, Cader T, Darby T. A Framework for Data Center Energy Productivity: White Paper #13. Beaverton, OR, USA; 2008. Available from: https://www.greenbiz.com/sites/default/files/document/GreenGrid-Framework-Data-Center-Energy-Productivity.pdf [accessed: 29/06/2018].
[23] Behrendt F, Schafer M, Belusa T. Konzeptstudie zur Energie- und Ressourcenezienz im Betrieb von Rechenzentren: Studie zur Erfassung und Bewertung von innovativen Konzepten im Bereich der Anlagen-, Gebäude- und Systemtechnik bei Rechenzentren. Berlin, Germany; 2008.
[24] Spec. SPEC Power and Performance Methodology - SPECpower ssj2008: Standard Performance Evaluation. Gainesville, VA, USA; 2007. Available from: https://www.spec.org/power_ssj2008/docs/SPECpower-Methodology.pdf [accessed: 29/06/2018].
[25] Anderson D, Cader T, Darby T. A Framework for Data Center energy Productivity: White Paper #13. Beaverton, OR, USA; 2008. Available from: https://www.greenbiz.com/sites/default/files/document/GreenGrid-Framework-Data-Center-Energy-Productivity.pdf [accessed: 13/06/2018].
[26] Stanley R, Brill K., Koomey J. Four Metrics Define Data Center "Greenness": Enabling users to quantify energy consumption initiatives for environmental sustainability and "bottom line" profitability - White Paper. Santa Fe, NM, USA; 2007.
[27] Kütz M. Kennzahlen in der IT. Heidelberg, Germany: dpunkt.verlag; 2011.
[28] Wilkens M, Drenkelfort G, Dittmar L. Bewertung von Kennzahlen und Kennzahlensystemen zur Beschreibung der Energieeffizienz von Rechenzentren. Berlin, Germany; 2013. Available from: https://depositonce.tu-berlin.de/bitstream/11303/3544/1/Dokument_13.pdf [accessed: 13/06/2018].
[29] UK Department for Environment, Food and Rural Affairs. Environmental Key Performance Indicators: Reporting guidelines for UK business. London, UK; 2006. Available from: www.defra.gov.uk [accessed: 29/06/2018].
[30] Perotto E, Canziani R, Marchesi R, Butelli, P. Environmental performance, indicators, and measurement uncertainty in EMS context: A case study. Journal of Cleaner Production. 2008; Vol. 16, N° 4:517–530.
[31] Dias-Sardinha I., Reijnders L. Environmental Performance Evaluation and Sustainability Performance Evaluation of Organizations: An Evolutionary Framework: Corporate Social Responsibility and Environmental Management. Corporate Social Responsibility and Environmental Management. 2001; Vol. 8, N° 2:71–79. Available from: https://onlinelibrary.wiley.com/doi/pdf/10.1002/ema.152 [accessed: 29/06/2018].
[32] Umweltbundesamt. Indikatoren / Kennzahlen für den Rohstoffverbrauch im Rahmen der Nachhaltigkeitsdiskussion. Dessau-Roßlau, Germany; 2012.
[33] Whitehead B, Andrews D, Shah A, Maidment, G. Assessing the Environmental Impact of Data Centres part 1: Background, energy use and metrics. Building and Environment. 2014; Vol.82:151–159.
[34] Haas J, Monroe M, Pflueger J. Proxy proposals for measuring data center productivity: White Paper #18. Beaverton, OR, USA; 2009. Available from: https://www.scribd.com/document/60185825/White-Paper-18-Proxies-Proposals-for-Measuring-Data-Center-Efficiency [accessed: 13/06/2018].
[35] Green IT Promotion Council. New Data Center Energy Efficiency Evaluation Index DPPE (Datacenter Performance per Energy) Measurement Guidelines. Tokyo, Japan; 2012. Available from: http: //www.greenIT-pc.jp [accessed: 13/06/2018].
[36] BITKOM. Prozesse und KPI für Rechenzentren: Leitfaden Version 1.0. Berlin, Germany. Available from: https://www.bitkom.org/Bitkom/Publikationen/Prozesse-und-KPI-fuer-Rechenzentren.html [accessed: 29/06/2018].
[37] Belady C, Patterson M. Green Grid Productivity Indicator: White Paper #15. Beaverton, OR, USA; 2008. Available from: https://www.thegreengrid.org/en/resources/library-and-tools/395-WP [accessed: 29/06/2018].
[38] Patterson M, Tschudi B, Vangeet O. ERE: A Metric for Measuring the Benefit of Reuse Energy from a Data Center: White Paper #29. Beaverton, OR, USA. Available from: https://eehpcwg.llnl.gov/documents/infra/06_energyreuseefficiencymetric.pdf [accessed: 13/06/2018].
[39] Green IT Promotion Council. DPPE: Holistic Framework for Data Center Energy Efficiency. Tokyo, Japan: [publisher unknown]; 2012.
[40] Aravanis A. Metrics for Assessing Flexibility and Sustainability of Next Generation Data Centers. In: 2015 IEEE Globecom Workshops; p. 1–6.
[41] Brown E. Electronics Disposal Efficiency: An IT Recycling Metric for Enterprises and Data Centers: White Paper #53. Beaverton, OR, USA; 2012. Available from: https://www.thegreengrid.org/en/resources/library-and-tools/235-Electronics-Disposal-Efficiency-%28EDE%29%3A-An-IT-Recycling-Metric-for-Enterprises-and-Data-Centers- [accessed: 13/06/2018].
[42] Patterson M, Azevedo D, Belady C. Water Usage Effectiveness (WUE) - A Green Grid Data Center Sustainability Metric. White Paper #29, The Green Grid, Beaverton, O, USA. Beaverton, OR, USA; 2011 [accessed: 13/06/2018]. Available from: http://www.thegreengrid.org/~/media/WhitePapers/WUE_v1.pdf.

[43] Cook A. Technology Carbon Efficiency (TCE): A Meter for Measuring Data Center Green Energy Impact - White Paper. New York, NY, USA; 2007. Available from: www.cstechnology.com [accessed: 13/06/2018].
[44] Bekas C, Curioni A. A New Energy Aware Performance Metric. Computer Science - Research and Development. 2010; Vol. 25, N° 3:187–195.
[45] Belady C, Rawson A, Pfleuger J. Green Grid Data Center Power Efficiency Metrics: PUE and DCIE. Beaverton, OR, USA; 2008. Available from: http://www.premiersolutionsco.com/wp-content/uploads/TGG_Data_Center_Power_Efficiency_Metrics_PUE_and_DCiE.pdf [accessed: 29/06/2018].
[46] Anderson D, Cader T, Darby T. A Framework for Data Center Energy Productivity: White Paper #13. Beaverton, OR, USA; 2008. Available from: https://www.greenbiz.com/sites/default/files/document/GreenGrid-Framework-Data-Center-Energy-Productivity.pdf [accessed: 13/06/2018].
[47] Blackburn M. The Green Grid Data Center Compute Efficiency Metric: DCcE: White Paper #34. Beaverton, OR, USA; 2010. Available from: https://www.thegreengrid.org/en/resources/library-and-tools/240-WP [accessed: 13/06/2018].
[48] Azevedo D, Patterson M, Pouchet J. Carbon Usage Effectiveness (CUE): A green grid data center sustainability metric - White Paper #32. Beaverton, OR, USA. Available from: http://tmp2014.airatwork.com/wp-content/uploads/The-Green-Grid-White-Paper-32-CUE-Usage-Guidelines.pdf [accessed: 13/06/2018].
[49] Pakbaznia A, Pedram M. Minimizing Data Center Cooling and Server Power Costs. In: ACM/IEEE Proceedings on the International Symposium on Low Power Electronics and Design; 2009; p. 145–150.
[50] Avelar V, Azevedo D, French A. PUE: A Comprehensive Examination of the Metric: White Paper #49. Beaverton, OR, USA; 2012. Available from: https://datacenters.lbl.gov/sites/all/files/WP49-PUE%20A%20Comprehensive%20Examination%20of%20the%20Metric_v6.pdf [accessed: 29/06/2018].
[51] Kaplan J, Forrest W, Kindler N. Revolutionizing Data Center Energy Efficiency. New York, NY, USA: McKinsey und Company; 2008.
[52] Spec. SPEC's Benchmarks.: Standard Performance Evaluation Corporation. Available from: https://www.spec.org/benchmarks.html#power [accessed: 13/06/2018].
[53] Avelar V, Azevedo D, French A. PUE: A Comprehensive Examination of the Metric: White Paper #49. Beaverton, OR, USA; 2012. Available from: https://datacenters.lbl.gov/sites/all/files/WP49-PUE%20A%20Comprehensive%20Examination%20of%20the%20Metric_v6.pdf [accessed: 16/06/2018].
[54] Deutsche Energie-Agentur. Leistung steigern, Kosten senken: Energieeffizienz im Rechenzentrum: Ein Leitfaden für Geschäftsführer und IT-Verantwortliche. Berlin, Germany; 2012. Available from: https://www.dena.de/en/home/ [accessed: 29/06/2018].
[55] Shah A, Bash C, Sharma R., Christian T, Watson B, Patel C. The environmental footprint of data centers. In: American Society of Mechanical Engineers: 2009 InterPACK Conference collocated with the ASME 2009 Summer Heat Transfer Conference and the ASME 2009 3rd International Conference on Energy Sustainability; 2009; p. 653–662.
[56] Umweltbundesamt. Nachhaltige Entwicklung in Deutschland. Berlin, Germany: Umweltbundesamt; 2002.
[57] Wang L, Khan S. Review of Performance Metrics for Green Data Centers: A Taxonomy Study. Journal of Supercomputer. 2013; Vol. 63:639–656. Available from: https://link.springer.com/content/pdf/10.1007%2Fs11227-011-0704-3.pdf [accessed: 13/06/2018]
[58] Marwah M, Maciel P, Shah A, Sharma R, Christian T, Almeida V, Araujo C, Souza E, Callou G, Silva B, Galdino S, Pires J. Quantifying the Sustainability Impact of Data Center Availability. ACM SIGMETRICS Performance Evaluation. 2010; Vol. 37, N° 4:64–68. Available from: https://dl.acm.org/citation.cfm?id=1773405 [accessed: 13/06/2018].
[59] ISO. ISO 14064: Environmental Management: Life Cycle Assessment - Principles and Framework. London, UK: British Standards Institution; 2006.
[60] Aggar M, Banks M, Dietrich J. Data Centre Life Cycle Assessment Guidelines: White Paper #45. Beaverton, OR, USA; 2012. Available from: https://www.thegreengrid.org/en/resources/library-and-tools/236-Data-Center-Life-Cycle-Assessment-Guidelines [accessed: 29/06/2018].
[61] Barroso L, Clidaras J, Hölzle U. The Datacenter as a Computer: An Introduction to the Design of Warehouse-Scale Machines. California, USA: Morgan & Claypool; 2009.
[62] Schödwell B, Wilkens M, Erek K, Zarnekow R. Towards a holistic Multi-Level Green Performance Indicator Framework (GPIF) to improve the Energy Efficiency of Data Center Operation -A Resource Usage-Based Approach. In: IEEE: Electronics Goes Green 2012+(EGG); 2012; p. 1–6.
[63] European Commission - Joint Research Centre - Institute for Environment and Sustainability. International Reference Life Cycle Data System (ILCD) Handbook: General guide for Life Cycle Assessment. Luxembourg, Luxembourg: Publications Office of the European Union; 2010.
[64] Acton M, Bertoldi B, Booth J, Flucker S, Newcombe L, Rouyer A. 2017 Best Practice Guidelines for the EU Code of Conduct on Data Centre Energy Efficiency. [place unknown]; 2016. Available from: http://publications.jrc.ec.europa.eu/repository/bitstream/JRC104370/2017%20best%20practice%20guidelines%20v8.1.0%20final.pdf [accessed: 13/06/2018].
[65] American Society of Heating, Refrigerating and Air-Conditioning Engineers. Data Center Networking Equipment – Issues and Best Practices: Whitepaper prepared by ASHRAE Technical Committee (TC) 9.9. Atlanta, GA, USA; 2013.
[66] BITKOM. Reliable Data Centre: Guide. Berlin, Germany; 2013. Available from: https://www.bitkom.org/Bitkom/Publikationen/ [accessed: 29/06/2018].
[67] BITKOM. Planungshilfe betriebssicheres Rechenzentrum. Berlin, Germany; 2014. Available from: https://www.bitkom.org/Bitkom/Publikationen/ [accessed: 29/06/2018]
[68] European Commission. Study on the Review of the List of Critical Raw Materials. Brussels, Belgium; 2017. Available from: https://publications.europa.eu/en/publication-detail/-/publication/08fdab5f-9766-11e7-b92d-01aa75ed71a1/language-en [accessed: 29/06/2018].
[69] Szczepaniak K. Das Rohstoffpotenzial von Rechenzentren: Quantifizierung kritischer Rohstoffe im Rechenzentrumsinventar zur Ermittlung des Grauen-Energie-Verbrauchs: Masterarbeit. Hamburg, Germany: Technische Universität Hamburg; 2018.

[70] Spec. SPECpower_ssj2008: Dell Inc. PowerEdge M710HD. Available from: https://www.spec.org/power_ssj2008/results/res2011q3/power_ssj2008-20110901-00396.html [accessed: 13/06/2018].
[71] Shehabi A, Smith S, Sartor D. United States Data Center Energy Usage Report. Berkeley, CA, USA; 2016 [accessed: 29/06/2018]. Available from: https://www.osti.gov/biblio/1372902-united-states-data-center-energy-usage-report.
[72] The E-Server Consortium. Energy Efficient Servers in Europe: Energy consumption, saving potentials and measures to support market development for energy efficient solutions. Brussels, Belgium; 2008. Available from: https://ec.europa.eu/energy/intelligent/projects/sites/iee-projects/files/projects/documents/e-server_e_server_final_publishable_report_en.pdf [accessed: 13/06/2018].
[73] Rasmussen N, Spitaels J. A Quantitative Comparison of High Efficiency AC vs. DC Power Distribution for Data Centers: White Paper 127. Schneider Electric; 2012. Available from: http://www.apc.com/salestools/NRAN-76TTJY/NRAN-76TTJY_R4_EN.pdf [accessed: 29/06/18].
[74] Bundesministerium für Umwelt, Naturschutz und nukleare Sicherheit. Energieeffiziente Rechenzentren Best-Practice: Beispiele aus Europa, USA und Asien. Berlin, Germany; 2009. Available from: https://www.borderstep.de [accessed: 29/06/2018].
[75] Oliveira F. Life Cycle Assessment of a High-Density Datacenter Cooling System. The TeliaSonera's "Green Room" Concept: Master of Science Thesis. Stockholm: Kungliga Tekniska Högskolan; 2012.
[76] Ramesh T, Prakash R, Shukla KK. Life Cycle Energy Analysis of Buildings: An overview. Energy and Buildings. 2010; Vol. 42, N° 10:1592–1600.
[77] Boehm M, Freundlieb M, Stolze C, Thomas O, Teuteberg F. Towards an Integrated Approach for Resource-Efficiency in Server Rooms and Data Centers. In: European Conference on Information Systems 2011 Proceedings; p. 1–13.
[78] Newcombe L, Limbuwala Z, Latham P, Smith V. Data centre Fixed to Variable Energy Ratio metric DC-FVER: An alternative to useful work metrics which focuses operators on eliminating fixed energy consumption. Swindon, UK; 2012. Available from: https://www.bcs.org/upload/pdf/dc_fver_metric_v1.0.pdf [accessed: 29/06/2018].
[79] United Nations Environment Program. Recycling Rates of Metals - A Status Report: A Report of the Working Group on the Global Metals Flows to the International Resource Panel. Paris, France; 2011.
[80] Hagelüken C. Recycling of (Critical) Metals. In: Gunn, G.: Critical Metals Handbook; 2014.
[81] Umweltbundesamt. Lebenszyklusbasierte Datenerhebung zu Umweltwirkungen des Cloudcomputing (Öko-Cloud-Computing). Dessau-Roßlau, Germany; 2017.

MIX
Papier aus verantwortungsvollen Quellen
Paper from responsible sources
FSC® C105338

If you have any concerns about our products,
you can contact us on
ProductSafety@springernature.com

In case Publisher is established outside the EU,
the EU authorized representative is:
Springer Nature Customer Service Center GmbH
Europaplatz 3, 69115 Heidelberg, Germany

Printed by Libri Plureos GmbH
in Hamburg, Germany